Blackhearts

Blackhearts

Ecology in Outback Australia

Richard Symanski

Yale University Press
New Haven and London

Copyright © 2000 by Yale University. All rights reserved.
This book may not be reproduced, in whole or in part, including
illustrations, in any form (beyond that copying permitted by Sections
107 and 108 of the U.S. Copyright Law and except by reviewers for
the public press), without written permission from the publishers.

All photographs are by the author. Designed by Nancy Ovedovitz and
set in Adobe Garamond type by The Composing Room of Michigan,
Inc. Printed in the United States of America by Sheridan Books,
Chelsea, Michigan.

Library of Congress Cataloging-in-Publication Data
Symanski, Richard.
Blackhearts: ecology in outback Australia / Richard Symanski.
p. cm.
Includes bibliographical references (p.).
ISBN 0-300-07819-6 (alk. paper)
1. Ornithology—Australia—Field work—Anecdotes. 2. Human
ecology—Australia—Anecdotes. 3. Zebra finch—Australia—
Anecdotes. 4. Symanski, Richard. I. Title.

QL673 .S98 2000
598'.099429—dc21   99-054708

A catalogue record for this book is available from the British Library.

The paper in this book meets the guidelines for permanence and
durability of the Committee on Production Guidelines for Book
Longevity of the Council on Library Resources.

10   9   8   7   6   5   4   3   2   1

Contents

	Acknowledgments	VII
	Map	X
	Introduction	1
1	Choosing Family	9
2	The Search for Breeding Redbeaks	30
3	Mysterious Behavior	64
4	The Gulf Widens	83
5	Business as Usual	129
6	Illinois Aftermath	193
	Postscript	202
	References	209
	Index	213

Photographs follow pages 46 and 110

*For Nancy, who knows, and
Cole, who doesn't—yet*

Acknowledgments

My wife, Nancy Burley, was sufficiently traumatized by the events narrated here that she tried to dissuade me from setting down this story. But I had decided to write the book before I had any idea how the adventure would unfold, and I would not be deterred by memories of all that I too would rather forget. Once the manuscript existed in draft, Nancy helped immensely—and willingly. She reread the pages almost as many times as I did, and we repeatedly went over the smallest details to ensure that I had been as fair as possible to the people about whom I had written. So in the last instance, if not in the first, it is Nancy who deserves the lion's share of credit for making *Blackhearts* better than it otherwise would have been.

Patty Gowaty, a colleague of Nancy's and a friend to both of us, gave me the important lead that directed me to Yale University Press. She also championed not only what I had to say but also how I said it. At Yale I received assistance and solid support from Jean Thomson Black, senior editor. There I also had the very good fortune to have the book meticulously edited by Vivian Wheeler. She tightened my sometimes loose and informal prose, and she showed me some mistakes in usage that I have made, I think, since the beginning of time.

When, at one point, it looked as though *Blackhearts* had run into some formidable snags, I was reinvigorated by encouragement, and good advice, from a longtime dear friend whom I have never met in person: Peter Gould. He persuaded me to stay with the project, even though at one point—finished as it was—the manuscript was about to be consigned to one of my black file cabinets.

Finally, Alan and Roz Andrews at Newry Station, where the finch research was carried out and where most of the events related here took place, were gracious in allowing Nancy and me and the students to live on the station, use their facilities, and roam at will in search of finches. Without their warmhearted generosity, this research would not have been possible and this tale could not have been told.

I give you this story.
This proper, true story.
People can listen.
I'm telling you while you've got time . . .
time for you to make something,
you know . . .

—Big Bill Neidjie, *Kakadu Man,* 1985

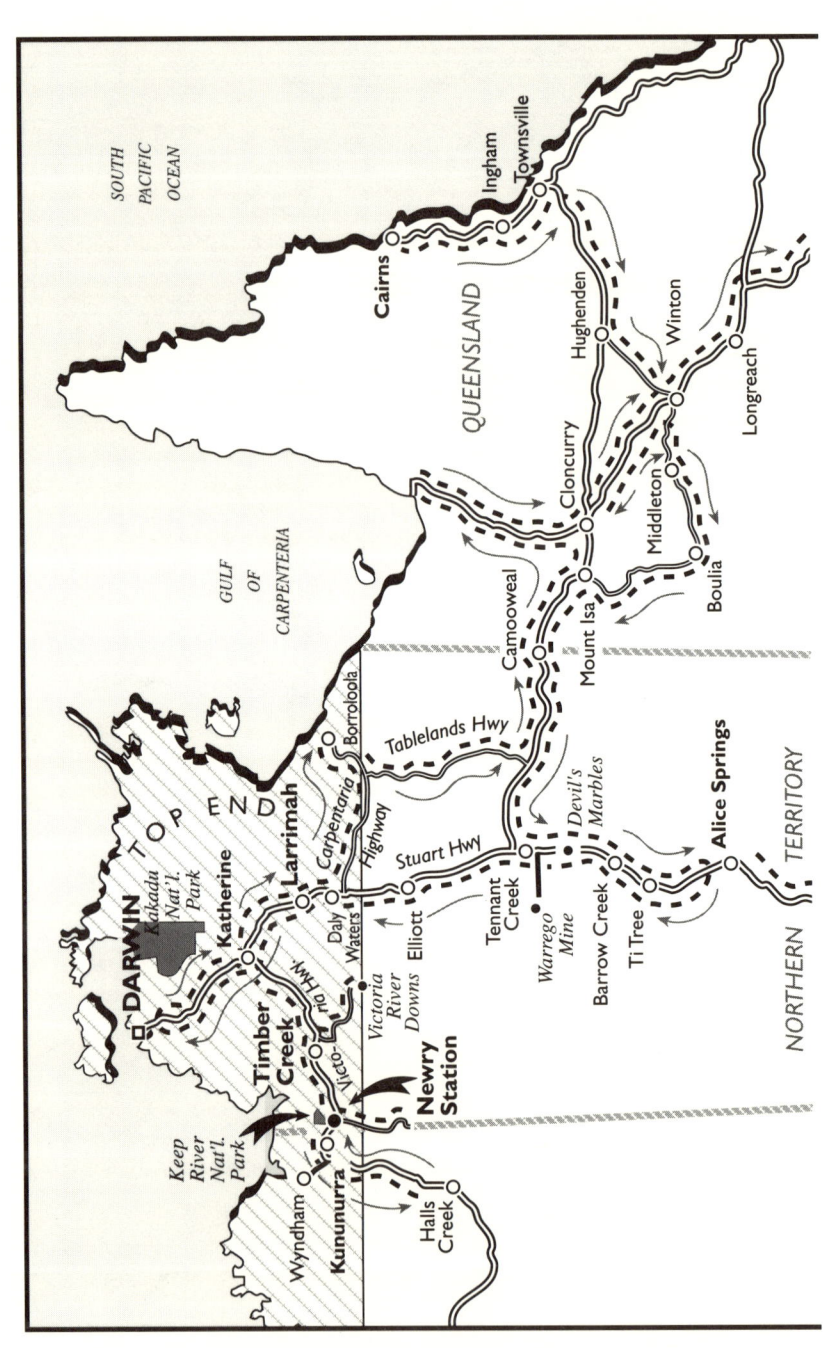

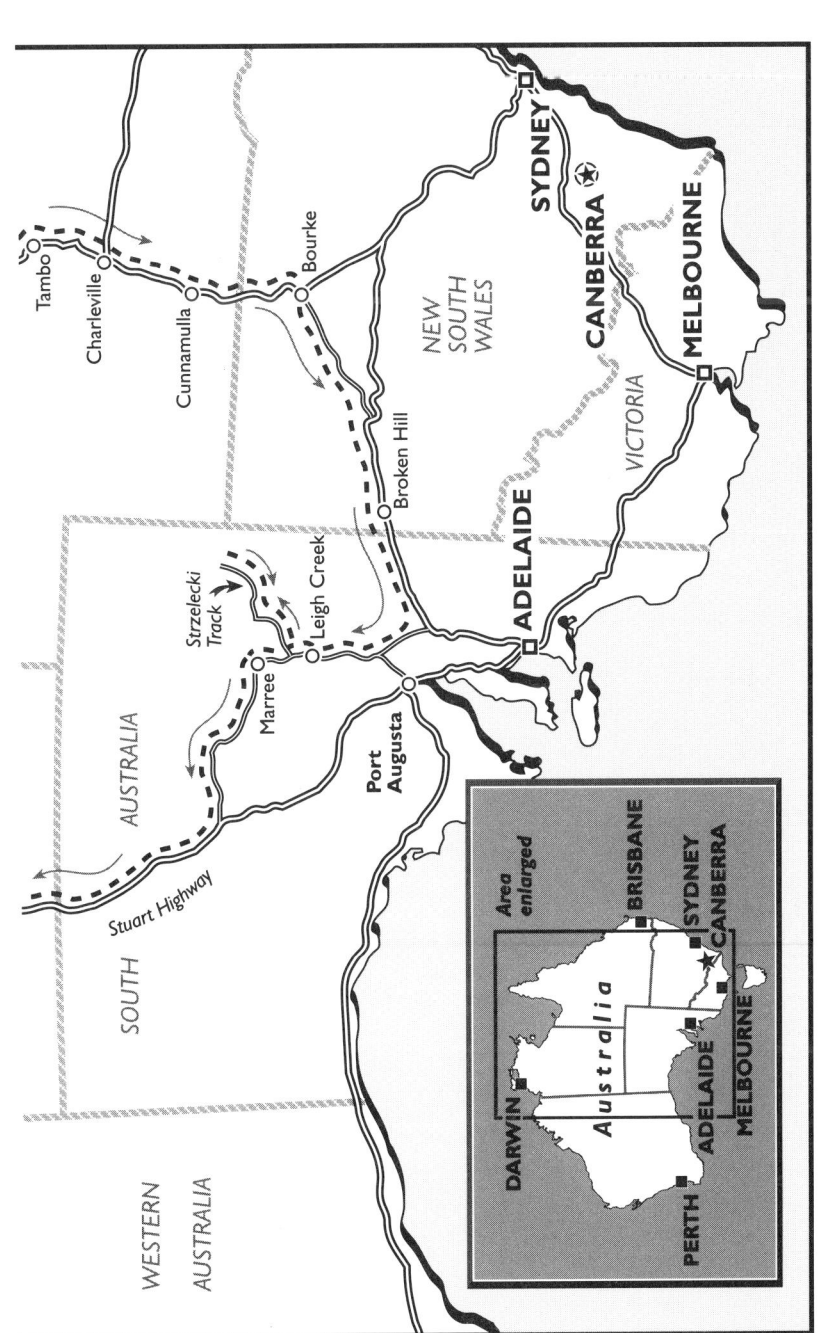

Itinerary of author and study field sites.

Introduction

In the spring of 1989 Nancy Burley, an evolutionary biologist and professor at the University of Illinois and my wife of many years, received a three-year grant from the National Science Foundation (NSF) to continue her long-term research on the social behavior of zebra finches. For more than a decade she had conducted laboratory experiments on the finches, with some quite spectacular results. To the surprise of scores of professional biologists, she had demonstrated that these tiny creatures, the third most popular cage birds in the world, are amazingly well tuned to evolutionary imperatives. When their appearance is manipulated by adding colored leg bands, they choose mates that are the most attractive—those that exaggerate familiar colors (red on males, black on females)—and then behave as if such attractiveness confers genetic advantage. In a species that is supposedly monogamous, more attractive males and females (as measured by these pseudomutations, the leg bands) contribute less to raising offspring; they manipulate the sex ratio of their young to favor the more "attractive" sex, and attractive males have more opportunities to philander than do unattractive males.

Until 1986 all these findings, and a number of others almost as notable, were achieved in laboratory aviaries. In the minds of some biologists this fact raised the question of whether the results were artificial, artifacts of unlimited food and water and the absence of predators, to say nothing of dozens of other factors that make indoor living space so different from nature.

In 1986–87 Nancy took a sabbatical from the University of Illinois and went to

Australia to investigate zebra finches in their homeland. Given the complexity of her research, she knew that it would be impossible to directly corroborate any major laboratory findings in a year or two. Indeed, were it ever possible to do so—which is doubtful—such a study would require a decade or more, and millions of dollars. Nevertheless, Nancy was able to demonstrate indirectly that her results were unlikely to be products of a counterfeit setting. In northern and southern and central Australia, she found sex ratios virtually indistinguishable from those recorded in her indoor aviaries. She verified that the beak colors of zebra finches, and the way they change with age and breeding season, are similar in the two environments. And she established that when zebra finches are taken from their native habitats and put in experimental chambers like those in her lab, the results are the same. These kinds of findings, reflected in a body of published research in the best journals of evolutionary biology (see References), convinced most of Nancy's critics that what she had discovered had to be taken seriously.

Still, not everyone was satisfied. Therefore, the more Nancy could reinforce her laboratory work with field data, the more she would be able to quiet remaining skeptics. In many ways these doubters are an incorrigible lot, acutely enmeshed in deep-seated prejudices about how organismal research should be done. The most fanatic among them believe that anything not discovered in nature is positively erroneous, not worth consideration. But such critics are mistaken, for they entertain wayward epistemological assumptions about research in general and biological knowledge in particular. One is that what is pure and undisturbed about nature remains so in spite of human intrusions. Yet the mere presence of field-workers among a species changes behavior, in ways often difficult to enumerate or measure. A second mistaken assumption is that field biologists are perspicacious enough to understand precisely the species or system under study. In truth, living systems are so complex, and so many practical difficulties are involved in getting good field data on even commonplace questions, that it can be excruciatingly difficult to infer "truth" from complex data sets gathered in the field. Combining laboratory and fieldwork is, of course, the best of all possible solutions, although to a good many practicing professional field biologists this statement seems to be anything but obvious.

If one reason for returning to Australia was to complement and buttress laboratory findings, another was pragmatic: both Nancy's 1989 NSF research grant and the one preceding it, which included her Australian stint in 1986–87, were funded in large measure because they proposed fieldwork in addition to laboratory experiments.

In Nancy's 1989 NSF grant application, she proposed to determine the degree to which zebra finches engage in what evolutionary biologists call extra-pair copulations: having sex with an individual other than one's mate. More technically and accurately, she was interested in the phenomenon of extra-pair fertilization (EPF), or the instances in which extra-pair copulation leads to production of offspring. To study EPF, she would use a recently perfected molecular technology that is popularly referred to as DNA fingerprinting.

The proposed field research presented some technical challenges. Despite her previous work in Australia, Nancy did not have an identifiable population of zebra finches to which she could return. Zebra finches are itinerant, and while Nancy has banded thousands of finches in the wild, most have never been recaptured. Moreover, because she suspected that rates of extra-pair fertilization might vary greatly within a population as social and environmental conditions changed, she hoped to compare results from two populations living at different densities. She wanted to identify two sizable colonies of finches, bleed and color band adults for individual identification, and subsequently draw blood from young at nests attended by identified parents. She hoped to get blood from at least twenty families at both sites, which was ambitious considering the high rates of nest failure she had previously encountered. Based on our earlier field stint, and because zebra finches are found over most of the Australian continent and are thought to breed opportunistically as conditions permit, Nancy assumed that several appropriate sites could be located without too much trouble. She asked ecologists in northern, western, and central Australia to keep track of weather patterns, being particularly eager to avoid areas where drought had been recently reported.

The most promising reports came from ornithologists stationed in Darwin, who verified early in 1991 that excellent breeding conditions were developing in the northern portions of the zebra finch's range. There grasslands and open woodlands are wedged between the arid interior and the tropical north. In 1986 we had encountered small populations of zebra finches breeding in this large area, and we could reasonably expect to find birds breeding under a diversity of conditions in relatively close proximity.

Ever since our 1986–87 stay in Australia, Nancy and I had been trying to have a child. Nancy was at that point in her life—in her late thirties and well established in her profession—where having a child was of more importance than almost anything else. Infertility doctors proved to be surprisingly unhelpful. Rather than give up, Nancy approached the biology of her body with the same determination she has always taken toward her research. She collected a mountain of data; she carefully analyzed it from half a dozen different angles; then, still frustrated with the answers and the protocols she was getting from "experts" in some of the Midwest's best hospitals, she got fertility drugs from a local hospital and literally mixed her own brew. Within months she was pregnant, and after the longest nine months of our lives, we were the very proud parents of a beautiful son.

But this great fortune augmented Nancy's problem of finding appropriate field sites where zebra finches could be captured and bled. With son Cole now an integral part of our lives and at an age when he required a good deal of attention, it would not be easy to drive thousands of miles in outback Australia in search of breeding zebra finches. Unlike the field trip of 1986–87, Nancy wanted to take along students to assist her. It would make little sense to bring them to Australia before identifying one or more viable field sites.

Although I had entertained thoughts of doing research in highland Colombia while Nancy was in Australia, I was eager to help. I had followed Nancy's career closely since its origins in the early 1970s, and over the years I had been strongly supportive. Now I had not only the opportunity to help her make her laboratory results yet more unassailable, but also the incentive of being with our child.

We planned enthusiastically. Because both of us loved Australia, we knew that this would be a great adventure for everyone. We had thoroughly enjoyed our stay in central Australia in 1986–87 and had often talked of returning—at one point we even considered making a permanent home there. That yearlong research adventure had been totally fruitful for both of us. Nancy had added some persuasive field evidence to her fascinating story, and I had written a set of field-based essays on conservation and Aboriginal issues in outback Australia (Symanski 1990).

But Nancy and I were also acutely aware of how the Outback might be perceived by others less enthusiastic than we were. So before we came to a final decision on any of the students who were to accompany us, we wanted to warn them of what was involved and what we expected. We wanted to be up front about the dreadful food and the outrageous outback prices and the hordes of flies that love the corners of human eyes. We wanted to be straightforward about everything that might be disillusioning, and not have to respond later to charges that we had not told our assistants that fieldwork is often boring, dirty, exhausting, and sometimes fruitless.

We would, we decided, meet numerous times before leaving with those whom we chose. Nancy could teach them the techniques that would be used in the field;

I could brief them on what to expect in ways personal and human—the Aussie factor. We could use the get-togethers to iron out logistical problems, to work as a team accumulating the needed equipment. We would have the team to dinner at our home, so that the students could become familiar with one another, with us, with Cole.

For these kinds of reasons, we decided not to advertise nationally for field assistants (as biologists often do). Over the years, many of Nancy's students—both graduates and undergraduates—had expressed a keen desire to visit Australia. On a campus of some forty thousand students, we were certain we could find qualified and interested candidates. We had heard enough stories about field seasons gone awry—not because the plants or animals couldn't be found or weren't breeding, but because the field assistants were "unknowns" who were ill prepared, had incompatible personalities, and couldn't control their sex urges.

In the end we decided to take three research students, all of whom would receive salaries from Nancy's research funds. We also would take one student to help care for Cole. She would be paid from our personal resources. We decided that I would go to Australia two months in advance of everyone else—more than enough time, we reasoned, to find one or more field sites. From our previous experience in the Outback and our interactions with Australians, we knew that a man would have many more opportunities and much more success than a woman in approaching complete strangers and extracting information about local ecology and bird life. Moreover, Nancy had teaching obligations, she needed to get her lab crew working smoothly in advance of her departure (her laboratory of six hundred zebra finches would remain in operation while she was in Australia), and she needed to finish ordering and assembling field gear.

So it was that I left for Cairns, Queensland, in early March 1991. Over the preceding two decades I had traveled widely in the Americas. I was prepared for surprises, for the unknown, and for an abundance of small adventures—which I relish. But I was ill prepared for the parade of revelations and eye-opening events, occurrences as much social as scientific, that would take place over the next six months and follow us back to Champaign, Illinois. What happened during this period, a time when Cole was coming into his own and Nancy and I were learning to be parents while trying to do serious fieldwork, was in turn exhilarating, frustrating, and thoroughly disheartening. In every sense, the venture was unforgettable.

This 1991 field experience did not have a Hollywood ending. But Nancy and I were not easily defeated, and in the northern summer of 1992 we again returned to

Australia. That time, however, it was just the two of us and our two-year-old son. Our mission had changed dramatically. Our sole aim in 1992 was to capture 140 long-tailed finches from two quite distinct populations and return with them to California. This second chapter of our Australian adventure with long-tailed finches provided new and unexpected challenges. Yet with all that failed to go as planned in the second phase of our outback tale, the end result was about as close to a happy ending as we might have realistically expected.

There are surprisingly few published accounts of what goes on among professional biologists when they team up to do fieldwork. From reading published scientific papers, field reports, and all but a few books, one gets the impression that fieldwork is all business—formal relationships, and little else. If personal or social conflicts occur, they are of little consequence, they do not affect the research, and they are easily if not always amicably resolved. In actuality, the reverse is often the case. People disagree, they fight, they hold grudges, they figuratively if not literally get divorced. After all, biology in the field is no different than any other realm of scientific or humanistic inquiry; it is preeminently a social endeavor. It is defined as much by individual prejudice, by personal and social urges, and by needs of the moment as it is by specific hypotheses and a corpus of scientific knowledge. The facts and published findings of field biology are consummately social, and must necessarily be interpreted in that light.

Perhaps it is not surprising that *White Waters and Black,* an informal sociology of sorts about field biology in the Amazon in the 1920s, is an underground classic among hard-core field biologists (MacCreagh 1985). I suspect that the book is a classic less because of its insights into the social dynamic that prevailed during an extended trip taken by six "Scientific Savants from Harvard" than by virtue of how infrequently such books have been written. With a fact altered here, a place reimagined, or a substitution of names, field-workers can vicariously identify with descriptions of uncooperative colleagues, sloth and incompetence, conflict at almost every turn, and the occasional hilarity of the unexpected moment in the midst of a serious search for scientific truth. Nor can those who have worked closely with others in the field fail to remember ways both big and small in which one or more aspects of the research as originally envisioned were changed because of the human dynamic.

From the moment I agreed to join Nancy and help with the field research, I knew I would write a book about our field experience. My aim was to narrate the adventure, and in the process to see what I would discover about everyone, not least myself. I wanted to record and analyze how Nancy's field team interacted as individuals only partly defined as scientists. I wanted to detail how she and I went about doing field research. Since I have always done fieldwork alone, I had virtually no preconceptions about how the adventure would unfold. The few that I did have were mistaken.

Most of this book was written very shortly after the events took place. From the time we began recruiting the students I kept extensive notes, and in the field in Australia I was constantly going to my notebook or laptop computer. The dialogue in this book is based on actual conversations, not words that I have put in anyone's mouth.

From the earliest days of our contact with the students we would take to Australia, I told them that I planned to write a book on our shared experiences. All the students were aware that I had already published several books; while there was no guarantee that a book would result from the venture they were about to undertake, they were cognizant of the possibility.

At least one person who read an earlier version of this book wondered why the students' story is not being told—why, to put the matter somewhat differently, they are not being given "equal time." In fact, several years ago, after the adventure was history and most of the manuscript was in first draft, I wrote to the students and asked if they wanted to give their version of several key incidents. They either said no or did not answer my letters. *Blackhearts* is, quite obviously, my story of what happened. Like any first-person narrator—indeed, any author—I have my own biases, my own unique point of view. I emphasize certain elements and I give little or no attention to what others would dwell on. Nevertheless, I have made every effort to be completely fair to the students as I knew them.

Are these students unique, and is it my duty to claim that they are by virtue of my verdict on their behavior? I do not, in fact, know if the students who worked with us in Australia are unique in all important respects, or in those aspects of their behavior to which I draw attention, or whether they are representative of graduate students generally. Because no surveys have ever been made, no one else has answers to these questions. My sense, and Nancy's, is that based on the numerous stories we have heard about the field experiences of other biologists, misbehaving or errant students of the sort we encountered—peculiarities aside—are not uncommon.

To provide a degree of anonymity to the students who were an integral part of this story, I have not used their real names.

However intimately I was involved in Nancy's research and the social dynamic of our adventure, many of my thoughts and interests lay elsewhere. I was trained as a social scientist. My Ph.D. is in geography, and I have no degrees or formal training of any kind in biology. For many years I have had an unwavering interest in fiction and other modes of depicting human behavior. Thus the methods I use in relating this story (and it is very much a story) are those of the novelist, the journal keeper, the travel writer. They are not those of the scientist as normally understood, especially the kind of scientist who mistakenly believes that his science is objective, conceived and undertaken apart from the quotidian human condition.

No matter how hard I worked with the finches and helped with the design and execution of fieldwork, I reflexively gravitated to issues that few field biologists would probe: the plight of the Aboriginal peoples in our midst; life at the cattle station on which we worked; the larger social history of the northern region of Australia where we trapped finches—its xenophobia, its agricultural base, its frontier mentality. In looking for breeding colonies of zebra finches in central and northern Australia, and then wandering every which way within easy reach of our principal field site, I was continually fascinated by all the telling cultural and historical tics and traits that make remote outback Australia unique. Insofar, then, as this story is about the ecology of a small piece of northern Australia in the first years of the last decade of the twentieth century, the "ecology" is every bit as much human ecology as it is the ecology that preoccupies single-minded field biologists.

While writing this book, I kept returning to *Black Hearts* as the title I would use. Such was my frustration and anger with the students who are at the center of this story. But I have opted for the sunnier title *Blackhearts*. This is a common name in the Outback for long-tailed finches; it refers to their striking black breast patches. I often found myself using the term "blackheart" when referring to a long-tailed finch. And before the students arrived at our field site, during a time when my mind was filled with nothing but warm thoughts about how the students would behave and how our adventure would happily unfold, I named the crude metal building in which we would cook and eat and process data the Blackheart Hotel. In spite of all that went wrong, I remain enamored of both blackhearts and the place I named in their honor.

1
Choosing Family

Eddie had been in Nancy's graduate class on behavioral ecology the semester he arrived on campus. She had been impressed with his performance, and near the end of the term remarked to me that she would have liked to have him as a graduate student. But he chose to work with another member of the ecology and evolution department who worked in Alaska. After Eddie's first year of graduate courses, he spent a summer there collecting data on the diet and distribution of the vole.

Midway through his second year at the university, Eddie took the first of the department's major examinations required of all potential Ph.D. candidates. Nancy was on the examination committee and again was impressed with his promise. This time she thought that perhaps he was an exceptional student, as fine as any she had seen in the dozen years since she began teaching and advising graduate students.

One day in January 1989, Eddie came into Nancy's office and said that he no longer wanted to work with his present adviser. He was tired of looking at seeds in rodent intestines. He wanted to do something involving more fieldwork with live animals, and in an environment less barren than Alaska's windswept tundra. He found Nancy's research on zebra finches appealing. Would she be willing to be his adviser, let him join the other students in her lab, go to Australia for fieldwork?

When Nancy met with Eddie a week later, she told him that she wanted him to maintain his ties with his current adviser. She would be his co-adviser. She already had five doctoral candidates, several other students who were working on master's degrees, and a score of undergraduates who collected data on the breeding behavior of finches in her large laboratory. What she didn't say to Eddie was that she wanted to use her remaining energy and hormones to become pregnant.

Even before Eddie came to her office the first time, Nancy had planned to send a young woman named Marta to Australia in the northern summer of 1989. One of Marta's tasks would be to look for a field site at which to study zebra-finch reproductive behavior. It occurred to Nancy that it would be mutually profitable for Eddie and Marta to go together to Australia. They could see if they could find viable research topics, and determine whether or not they enjoyed fieldwork in the remote Outback.

Nancy arranged for Marta and Eddie to stay at Keep River National Park in the northwest corner of the Northern Territory. In 1986 Nancy and I had teamed up with Heidi Blake, an ecologist from the Conservation Commission of the Northern Territory. Heidi's current principal interest was the Territory's declining populations of Gouldian finches. Although the three of us had successfully trapped several kinds of finches at locations south and east of the park, Eddie and Marta had none of our good fortune. They found only long-tailed and masked finches—no zebra finches.

After several weeks of frustration and no hope of matters improving, Nancy told Eddie and Marta to take the long bus ride south to Alice Springs, to the Commonwealth Scientific and Industrial Research Organization (CSIRO), Australia's premier government research institution. Nancy had learned that it would be easy for Eddie and Marta to capture zebra finches there and watch them at their nests. From scientists she knew and respected, she had heard that the finches were so abundant at CSIRO that they were breeding on window ledges.

Eddie and Marta found a trailer to live in several miles from their field site and traveled back and forth on bicycles. This soon proved to be too taxing for Eddie. He was constantly tired. He was aware that he had a weak heart, but didn't think his body was telling him that this was the problem. Eddie had had Hodgkin's disease as a teenager; by his account, the doctors had given him far too much radiation, which had weakened his heart.

When Nancy learned of Eddie's health problems, she told him to return home or to seek medical advice and have his heart monitored. But Eddie had his own agenda. He wrote to Nancy and said that he probably only had the flu.

Once home, the verdict was harsh. Eddie was told that he needed a heart transplant and that if he didn't get one very soon he'd be dead. It was only a question of deciding whether to go to a hospital in Palo Alto, Pittsburgh, or St. Louis. Eddie reeling from Eddie's death. She's clearly the best in the department for our needs. And we agreed not to look elsewhere."

We had, indeed.

In the days and weeks that followed, I repeated my dissatisfactions, especially Jean's insistence that she wouldn't help with Cole and that she had to leave early for her sister's wedding. "Whom would you take instead?" was Nancy's stock response. I'd shake my head and start to speak, then remember that it wasn't my research nor was it my grant that would support the students. My take-charge, volunteer, no-strings-attached job was to take care of all the before and after and in-between logistics, choose a field site, and help with the research when needed. It was simple, I told myself. Nancy couldn't do all this on her own, especially now with Cole very much at the center of our lives.

Nancy has an immense appetite. Six hundred or seven hundred eager-to-breed zebra finches in her Illinois laboratory were barely sufficient for her research needs. Still, once, twice, sometimes three times a year, she'd reduce her always-growing population to keep the density at a reasonable level. A few finches were sold to local pet stores, some to wholesalers in Chicago. Many went to shifty, penny-pinching aficionados who bought and sold and traded birds at a farming community fairground midway between Champaign and Chicago. The small profits from the sales bought a few bags of finch seed, some cuttlebone, cleaning supplies.

By 1985 Nancy was increasingly swamped with university work. She was grateful when I agreed to take charge of disposing of her redundant finches. Usually I'd get her to ask some of her undergraduate lab students to accompany me: to help sell the birds, to get my mind off the tedious corn and soybean fields on the drive north, to find out what state universities were now calling good students. Most were eager to go; some went with me several times. For those who had grown up in a scrubbed suburb of Chicago, the homey fair and the down-home haggling were experiences not easily forgotten.

It was on one of these trips in the fall of 1989 that I met Nat. Like many of the undergraduates who worked in Nancy's lab, Nat collected data and made behavioral observations. She counted eggs and hatchlings, weighed birds, and took other measurements. She sat behind a one-way mirror with binoculars and a stop watch and recorded the sexual gambits of Nancy's color-banded finches.

Right away Nancy had seen that Nat was slow at what she did. And she was not much of a listener. Yet Nat had a trait that Nancy found singularly attractive: she was meticulous. When Nancy assigned her the task of checking the work of other students, Nat was quick to find errors. That was important to Nancy, for nothing is deadlier to good science than suspect data.

It was Nat's observations of two biologists who worked for the Illinois Natural History Survey that alerted me to her distinctiveness. I would not have paid much attention to what she said had I not followed the research and personally known those of whom she spoke. I was struck by the shrewdness of her observations, how much of what she said fit with what I knew about them. Nat had done some housecleaning for one of them. He was a known stickler for detail, someone never satisfied with the quantity or quality of his scientific data. Her canny characterization of his fanatic fastidiousness seemed as accurate as anything I would hear from a trained sociologist. In one of her stories, the scientist was alleged to have a bedroom closet in which lightbulbs of every imaginable size were scrupulously organized by wattage and regularly checked against a computer-based inventory of household goods.

Nat worked hard at making herself different. Like Jean, she wore a variant of the ecologist's field uniform. But she supplemented it with various kinds of ragtag clothing from army surplus stores. Her big brown eyes peered searchingly through thick-lensed granny glasses. Framed by a boyish haircut, Nat's square face and slight upper body gave the impression of an orphan, a waif, someone who needed protection from a threatening world. Offsetting these childlike qualities were the premature gray in her hair, the bifocal lines in her lenses, her strong opinions about everything. Nat refused to subscribe to commonplace beliefs. Her real name was Natalie, which she detested. It reminded her of Natalie Wood. She preferred the androgynous Nat.

Recently, behavioral ecologists have come to appreciate that birds long described as monogamous may not always be faithful to their mates. There are sound reasons why they shouldn't be. For males, philandering opportunities should increase their reproductive success at little cost, since males usually help rear the young of only their social mate. A principal way females can benefit from extra-pair matings is by increasing the quality of their offspring. If a female cannot become the social mate of a male with especially favorable genes (because he is mated to another female), she might nevertheless garner some of those genes for her offspring through extra-pair copulations (EPCs). A current theoretical debate among behavioral ecologists is whether EPCs are costly to females: do their social mates lower the degree of parental care if their paternity confidence is low? By the late 1980s this question had become framed as a black-and-white issue, with several mathematical modelers arguing that males should ignore the question of paternity when making decisions to give parental care. Nancy thought this reasoning simplistic for most birds. She anticipated that, as with most interesting ecological questions, detailed empirical studies would show that "truth" came in many shades of gray.

Mating behavior in birds is hard to observe under the best of circumstances. Not unlike humans, many avian species like privacy for their romantic moments. It is rather rare to catch a fluttering philanderer *in flagrante delicto*. DNA fingerprinting, in recent years a hot technique in evolutionary biology, has been called a panacea. Properly used, it is possible to document extramarital episodes by examining the similarity in DNA structure between "mother" or "father" and alleged offspring.

DNA fingerprinting as a method for proving genetic relatedness is not without its critics, and it is a technique not easily learned. Success requires not only mastery of a detailed protocol meticulously followed, but also careful judgment. DNA fingerprinting is one part science and some significant part art. Nancy's zebra-finch research was at a stage where she needed to know the extent to which particular individuals were unfaithful. She already had numerous observational data giving approximate rates of infidelity; what she now needed were rates of extra-pair fertilization (EPF).

The immediate reason was to examine the validity of her finding that band color in her colony of zebra finches affects reproductive success. Years earlier she had determined that the attractiveness of birds as mates could be altered by placing plastic bands of certain colors on their legs. Females found red-banded males attractive, those with green bands unattractive. In long-term breeding experiments on captive birds, red-banded males raised more offspring than green-banded ones. If the EPF rates were the same in the clutches reared by red-banded and green-banded males, then the generalization that attractive males had enhanced reproductive success was true. But what if attractive males, which Nancy had also found showed lower rates of parental care, fathered fewer of their socially reared young? Nancy believed this to be an unlikely possibility; indeed, she thought attractive males should have higher paternity confidence than unattractive birds.

There were other reasons why Nancy wanted to investigate this question. While some researchers had begun to report species-specific rates of EPF, Nancy felt that rates should vary within species as social and ecological circumstances varied. Her color-banding findings allowed her easily to manipulate social circumstances. Also, zebra finches breed under a variety of ecological conditions, with concomitant variation in synchrony of breeding and population density. She suspected that those factors would also affect EPF rates. At the time, most researchers had what Nancy would describe as a male-biased view of extra-pair matings. Nancy surmised that the color-band phenomenon would be useful to illustrate the occurrence of female tactics. Thus, for a variety of reasons, zebra finches would be well suited for studying the complexity of reproductive tactics displayed by males and females of a species that is putatively monogamous.

As soon as Nancy could attract the resources to set up a DNA lab, she delegated one of her graduate students, Fred Lance, to learn the technique. Fred was the first doctoral student Nancy had had who was pursuing a project on zebra finches. At the time she established her DNA lab, Fred was testing hypotheses about mate choice. He also had an interest in genetics, so it seemed important for him to know the real parentage of offspring. Fred was ambitious and a hard worker, and he understood that familiarity with molecular techniques would improve the likelihood of his getting a good academic job.

By the time Fred began to master the art and science of chemical mixing, he was well along on his dissertation. But he was simultaneously trying to handle the pressures that come with a family. Fred and his wife had just had their first child, and his wife was tired of the impoverished graduate-student life. She also wanted another child.

Fred located a postdoctoral position at a university in another state. The job required him to leave Nancy's laboratory and the university before completing the work he had planned, and well before he could write it up for publication. Knowing that the job would only prolong his quest for a degree and delay publishing the results of his zebra-finch research, Nancy advised Fred to stay at Illinois until he had the doctorate in hand. He ignored her advice and took the postdoctoral position.

Nancy saw that Fred's premature departure could create a major bottleneck in her research. Unless someone could pick up where Fred had left off, she would not only have a backlog of unfinished research papers, but would have to explain to the National Science Foundation why she had failed to come up with results, as she had contractually agreed to do.

As soon as Nancy elected to set up a DNA fingerprinting lab, she made a firm decision not to get intimately involved in the process. She did not want to expose a potential fetus to chemicals with unknown mutagenic and carcinogenic properties, certainly not on top of the fertility drugs with which she was experimenting. So Nancy needed someone to pick up where Fred would leave off—ideally before he left the university. However, there were no obvious candidates in Nancy's lab or on the horizon, and she had become increasingly wary of taking on anyone who lacked outstanding credentials. Her predicament was complicated by the impending trip to Australia. She had no money to train and hire an outside technician to do the fingerprinting, nor was she sure this would be a good idea. She wanted to find a person who would have an investment in the technique, who would commit to staying with the task to completion—someone who would not only fingerprint the zebra-finch blood in the lab, but also process the blood from the field trip to Australia. Someone to both do the DNA work and assist with the Australian fieldwork might be the best of all possible solutions. But who?

Nancy had taken on a new, bright-eyed Ph.D. student who would begin in the fall of 1990. Still, as with virtually all first-year graduate students, it was impossible to evaluate real interests, strengths, weaknesses, or a long list of crucial intangibles. Despite what the student had told Nancy and me in a captivating interview the previous winter, once on campus she might not want to work with zebra finches. She might have no interest in toiling in a DNA lab with nasty chemicals. Nancy and I concluded that her new student came in at no better than forty-to-one in terms of coping with showerless days, erratic mail deliveries, and outback telephones that unaccountably go dead in the middle of a conversation.

After considerable discussion, Nancy and I agreed that the person who might best fit our needs was Nat. We had heard numerous stories about her self-initiated wanderings in the Shawnee National Forest in search of cowbird eggs. Nat was universally declared by those involved to be among the very best at finding bird nests. Based on her work in Nancy's lab, it was also clear that Nat was very careful when collecting data. From this we inferred that she'd be equally attentive to the picky protocols and fragile chemicals that result in a successful DNA lab.

Still, Nat had some obvious liabilities. She did not seem to have much interest in pursuing a Ph.D. In looking over Nat's university record, Nancy discovered that her grades were poor. Nancy made the point with administrators in the admissions office that Nat's experience to date in Nancy's laboratory, and her high marks for work in the Shawnee National Forest, showed that her grades were not a true measure of her abilities. She made a convincing case on Nat's behalf.

Nancy called Nat and asked if she were interested in pursuing a master's degree.

"I don't know," she said. "Why do you ask?"

"I'm looking for another person to take to Australia. I also need someone to take over my DNA fingerprinting lab. Based on your work with the zebra finches, you would seem to be well suited for the job."

"Oh! I sure am interested in going to Australia. And I know I can do the DNA fingerprinting if someone shows me how."

Nancy said that, before he left, Fred would teach her everything he knew about the technique. "You'll then be expected to finish the work that Fred has started on lab birds. I'll want you to do whatever fingerprinting is required on finches bled in Australia. For your thesis you can use fingerprinting data generated from either the lab birds that are already bled or from those that you'll help catch and bleed in Australia."

"Oh, that's great! I'll tell you now, I'll go."

Nat soon let the whole world know that she was bound for outback Australia. Except for a week-long car trip deep into the bowels of wintry Texas, she had never been out of Illinois. She would soon be twenty-six and a university graduate, and she still hadn't been on an airplane. Not only would Australia be the first real adventure of her life, but it offered her several bonuses. The one dearest to her heart was the chance to find herself on the continent with seven of the world's twelve deadliest snakes.

"God, that'll be something, being around all those snakes!" she often exclaimed.

Copperheads and garter snakes and anything else that resembled a limbless reptile enchanted and inspired Nat. Now she'd have the chance of a lifetime to see up close Australia's renowned and deadly king brown snake, and who knew what else she'd come upon?

Not long after Nat began working alongside Fred Lance, I expressed my concern to her about the seriousness of her pledge to do the DNA fingerprinting. For a variety of personal reasons I'd become skeptical of verbal commitments and, as was my habit, I often saw Nancy's best interests as my own.

Nat wasn't happy when I reminded her of the commitment. She said merely, "Nancy already told me what the deal is."

Initially, we had planned to take two students to Australia. But even before we asked Jean and Nat to accompany us, we made budget calculations for several different scenarios. We decided that relative to the cost—even though some of it would come out of our pockets—it would be worthwhile to bring a third student. It had been five years since Nancy's last trip to Australia, and it could be another five before the next one. Another pair of young hands and legs and an active mind would allow Nancy to increase sample sizes, which was often the difference between acceptance and rejection of an arresting hypothesis. Another student would make it possible to collect data on questions that had not occurred to Nancy prior to setting foot in the field, data that might well lay the groundwork for future grant proposals. It is no secret that because of the intense competition for federal grants, preliminary data (and a reasonable sense of what one might find) are prerequisites for funding.

Nancy approached a couple of senior biology majors who seemed to have considerable research promise. Neither had any interest in going to Australia. Too grubby, too many early mornings, not enough play time, each reasoned in his own way.

Some months before Nancy talked to these students, she was approached by an unfamiliar undergraduate in her department. Tim had taken none of Nancy's classes and had not worked in her laboratory. But based on a recommendation from a colleague and personal friend, Lana Otsinger, and from another professor whom Nancy knew only by reputation, Tim sheepishly asked Nancy if she would help him with research he was doing on bats in Costa Rica. Tim confessed that he was desperate. He said that for more than a year he had been working with one of Nancy's colleagues, but now they were estranged.

Nancy phoned the person who had worked with Tim. She also called a few friends and colleagues elsewhere on campus who might know something about him. She learned little. One source said that he had been "too independent." Another said that his understanding of bat behavior and ecology was "out of synch with received wisdom." Someone else confided that Tim had put his hands where they didn't belong, on some files in an off-limits cabinet in a faculty office.

Nancy weighed the information she had received, especially the sterling recommendation from Lana Otsinger. Both of us considered Lana to be among our closest friends. We often had dinner with her at our house or hers. Unmarried and with no children of her own, she loved being around Cole. Lana was someone we trusted, who understood our needs and what we would expect from a student. She left no doubt in our minds that Tim had been unfairly maligned, and that he would be an excellent choice to take to Australia.

Beyond the faith that we put in Lana's assessment of Tim, and what Nancy had learned from others, she had no interest whatever in playing the role of professorial detective. She told Tim that whatever had happened was now history; she would give him a fresh start.

Nancy was quickly drawn in by Tim's enthusiasm, his interest in research and things new. Every time he came to her office he had lists of questions. She found that he followed instructions well and was as efficient as the best of her graduate students at returning promptly with analyzed data and clean drafts of what he had written.

Tim had, it seemed to Nancy, unbounded energy, an insatiable appetite for work. Besides carrying a full menu of memory courses, he was being tutored by a professor in another biology department on state-of-the-art honey-bee research. He pestered Nancy to show him how she went about getting novel insights into the sexual behavior of zebra finches. She found enough money to pay him to work in her lab, so that he could learn the methods she used in data collection. And she helped him get money for a return trip to Costa Rica at Christmas for more exploratory research on bats. Tim seemed a mature twenty-one.

Nancy painted an impressive picture of Tim for my benefit. She said that he must have worked almost nonstop during his summer and Christmas stints in Costa Rica, he had returned with so much data on a difficult-to-study species. "He offers something special if we take him to Australia," she said. "Since he's planning to travel for a year after he graduates, he would stay with us into the fall of 1991 if we needed him."

It seemed unlikely to me that we would have to spend more than four months in the field to get the data Nancy needed. But who could know in advance what her needs might be, what might go wrong? The first trip to Australia had taught us that zebra finches were a lot harder to catch at the nest than Nancy had been led to believe in advance by Australian biologists.

"Have him meet me at your office on Friday afternoon," I said to Nancy. "I'll take him over to the Nineteenth Hole and see what I think."

I poured the first beers and then, for Tim's benefit, spread a huge map of cattle stations in The Northern Territory on the table. I described the vastness of the Outback, told some colorful stories about station life, how and why the Conservation Commission was shooting wild donkeys, horses, pigs. Tim's questions were endless. He had boyish good looks, a thick mane of black hair, a solid frame, a disarming lack of affectation. His enthusiasm seemed boundless.

I had decided in advance not to ask Tim about his undeclared and unexplained war with one of Nancy's young colleagues. Everyone needs new beginnings, I reasoned. Wasn't I always looking for ways to erase pieces of my own history? While we talked over our beer, I concluded that Tim was a student who would keep me on my toes, not let my mind atrophy from the boredom of collecting data, or from too many hours under an oppressive tropical sun.

We had planned to bring the "Oz family" together for regular meetings when university classes began in the fall. But our plans were derailed because Jean was taking every single day that I'd given her to decide whether or not to go with us. On the evening of October 15, the last date I'd given her, I called her and asked, "Are you still interested in going to Australia with us?"

"Let's go!" she said.

We began meeting regularly. We gathered at the White Horse Inn, a restaurant and bar famous with university students, where Nancy and I sometimes ate crispy chips and hot salsa and called it dinner. Or we met at our old two-story home (I had once thought it would be fun to turn the long living room into a finch flyway). Or we gathered in Nancy's spacious new $200,000 lab, in a small room with white walls and an oversized poster of a beluga whale. There thousands of zebra finches had been held by their necks while Nancy and scores of students matched the colors of their beaks against numerically coded Munsell color chips. There Eddie had carried out foraging experiments on zebra finches after Nancy became his co-adviser.

We met once specifically to discuss a twenty-point contract that Nancy and I had put together and wanted everyone to sign. We wanted to avoid misunderstandings. The contract specified that Nancy would pay round-trip airfares and all in-country travel that was research related. She would pay salaries, about what University of Illinois graduate students were making as teaching assistants in her department. Nancy would pay for all field and cooking supplies, virtually everything the students would need save bush hats, designer shorts, and food. To relieve the students of unnecessary burdens, keep their minds free for fieldwork, and their nonresearch time free for doing whatever they fancied, I agreed to do all the banking, keep a set of always-open books, and give them advances against their salaries anytime they needed extra money.

The contract spelled out various contingencies. If we didn't get along with the students, or Nancy and I were unhappy with the work they were doing and nothing could be worked out amicably, we had the option of asking them to leave the field site. Should a student depart earlier than initially agreed upon (other than for a reason such as sickness in the family or personal illness), the student would be asked to return half of the round-trip airfare. Jean's contract had an additional clause. She could return home a week before her sister's wedding.

In our open discussions with the students, we emphasized our desire to work together. We asked that everyone speak openly and candidly of problems as they arose. We made it clear that we did not want an environment in which anyone was pursuing hidden agendas, or for whatever reason festering with unspoken resentment.

With the initial meetings behind us, surprises began to appear. Tim's girlfriend, Amy, who we did not know existed when we asked Tim to join the team, started coming to the White Horse Inn meetings to be alongside her boyfriend. Like Tim, she was graduating in June. She wasn't eager to make immediate use of her degree in economics, and she began musing out loud about some traveling she wanted to do. Would it be all right if she came to our field site and spent some time with her boyfriend? she asked an imaginary person who inhabited a space somewhere between Nancy and me. She said she could get money from her father for the airfare to Australia. If we were generous enough to let her come to our field site, she'd help with our baby-sitting needs. She loved children. Her services would be free. What did Nancy and I think about that? Amy and Tim wondered aloud.

Nancy and I discussed the overture in private. Neither of us thought that Amy's presence at the field site was a particularly promising idea. We saw possible problems, and we imagined others.

Amy continued to come to our meetings, and the more we saw of her the more we liked her. She seemed sincere and unaffected. Only Nat was unsettled by Amy's presence. When Nancy asked about the source of her irritation with Amy, she said she didn't like "sorority types." That vague accusation didn't ring true with either of us. Amy didn't fit any sorority stereotype that we were aware of—no social snobbery was evident; she didn't even hint that she'd been president of her sorority.

As the weeks turned into months, Nat reluctantly admitted that Amy had several advantages and that she added a cheerful or humorous note to overly serious moments. And Amy obviously loved Cole. Whenever she was around him, she was warm and cuddly with him. She caught his eye. She made him smile and laugh. She enjoyed playing with him. This was a huge plus for me, and no less so for Nancy.

Nancy and I became realistic. Cole would be a lot of work, and he could pose a real problem if we caught a great many birds at one time. We could get everyone to help, but that might not be enough, I reminded Nancy. She agreed.

I called Amy and said she could come with us to Australia as Cole's nanny. We would pay her out of pocket for all her travel expenses once she landed in Australia.

"It's much more than I could have hoped for," she exclaimed. "Tim will be real happy."

Now she could stay in Australia until we no longer needed Tim. Afterward the two of them could wander off to see the Great Barrier Reef and Sydney and Melbourne, then head for Southeast Asia, India, and the Middle East. "On bread and cheese, just like we did in France and Greece two years ago," Amy said.

Soon Tim began asking for special help from us. Could we arrange the tickets this way . . . no, that way . . . no, wait a minute, I'm not sure just what the best plan is for going around the world. We want to see New Guinea and . . . We finally told Tim that he'd have to make his own arrangements. "But keep your departure date open-ended," Nancy said. "I may want you to help me into early fall, as you said you could."

We wanted to be fair, so when Jean said that her boyfriend, Alfred, might want to spend his summer vacation in Australia, I said, "Okay, if he doesn't interfere with the research and gets along with everyone, he's welcome to spend a couple of weeks with us."

"Great!" she said.

"Nancy and I would like to meet him before we leave."

"You'll love Alfred. He's trapped every kind of bird that lives in Illinois."

For whatever reasons, the invitation to Alfred was ignored.

Only Nat failed to come forth with the I've-got-a-special-friend-and-can't-do-without-him-for-the-summer-can-he-come line. Her boyfriend was too poor to pay for an expensive air ticket, and he didn't care that much about Australia. And lately he was no longer exactly a boyfriend, Nat let slip at one of our team meetings.

Gathered around a long table near the fireplace one afternoon at the White Horse Inn, we began to discuss each team member's strengths and special abilities. Nancy outlined the kinds of data she hoped to collect and the tasks all of us might be doing if I were able to find breeding zebra-finch colonies. From the beginning, Nancy had expected Jean to assume the role of leader among the students, and she had made this explicit to Jean. But on this very afternoon, before Nancy got around to the chain of command among the students, Jean said, "I'm looking forward to not having any responsibility, just following directions. I've been working hard at the Natural History Survey since Eddie died. It's going to be a great break."

Nancy looked over at me and furrowed her eyebrows.

Nat jumped in, as if she'd been waiting for an opportunity to assume the leadership mantle. "I've been doing lots of reading and have ideas about what we need to bring. I know what we need to do and talk about." Turning to Nancy, she began by saying, "Have you gotten a good snakebite kit yet? There are many poisonous snakes in Australia, you know."

"We'll be taking one, but we haven't gotten that far in our planning yet," Nancy said.

"What kind of tents are you going to buy? They have to have sturdy floors, and window flaps for good ventilation. If you tell me how much money you have to spend, I'll pick them out."

Nancy more or less repeated what she had just said.

"But who is going to arrange for our bus tickets once we arrive in Australia? Have you given any thought to where we're going to stay before we get to the field site?"

"We don't yet know where we'll be," I said. "I'll take care of those arrangements once I find the finches."

When we were able to get everyone to agree on a time, we began meeting on a Saturday or Sunday afternoon in Nancy's lab to learn various techniques for working with finches. Everyone was agreeable to just about any time Nancy suggested—with the exception of Jean. Some weekends she couldn't meet at all: she was too busy, or none of the times was convenient. Or she'd have to leave early because she had other plans. Alfred plans, everyone speculated when Jean remained vague.

Nancy showed us how to bleed finches under the wing. She showed us where to go for blood if there was an insufficient quantity in the wing veins, as sometimes is the case for young still in the nest. She made cutting the jugular vein and keeping the bird alive seem easy.

Nancy took everyone through the Munsell protocol, matching a bird's beak with scientifically coded color chips to get a value for hue, chroma, and brightness. Over the years Nancy had measured color on thousands of birds and taught scores of students the easily learned technique. It was a cakewalk as long as a person was not color-blind; then it was impossible.

Nat's Munsell scores were off scale, unlike anyone else's for the same bird beaks. Nancy had to tell her that she wouldn't be able to participate in the color measurements. Nat was crushed. Nancy consoled her; we told her that there were plenty of other valuable jobs for her in the field. "It's no big deal."

Each week there were new requests, small problems, recurrent issues. Nat was anxious because we hadn't yet bought the plane tickets. She had to see them to know that she was in fact going and that the whole project wasn't simply a dream, an empty promise. We told her to relax and did what we could to quiet her anxieties.

Tim continued to muse out loud about his world travel plans. He wanted to see and do everything, and he was putting off giving us a final departure date until the last minute.

Jean's principal preoccupation was wondering how many important Australian research-related contacts we could provide her with.

Nancy decided that it was imperative to take six homemade walk-in traps for catching zebra finches. They had worked well on our last trip to Australia. Even if we mist-netted birds, the walk-in traps could be handy. Some birds don't go to mist-nets. Because of light conditions or lack of water, we might not be able to use them. Once we got finches accustomed to walk-in traps, we might be able to catch fifty or a hundred in a couple of hours.

The lightweight trap frames were made in sections by men who worked in the university metal shop. Nancy bought huge rolls of plastic mesh. She taught us how to sew the mesh to the sections. Nat stitched them well enough to last fifty years. I sewed them inside out and upside down and left-handed—a true misfit. Jean was busy with personal matters and didn't sew any. Nancy's favorite doctoral student, Dave, volunteered to sew as many sections as she gave him. He did a better job than anyone. It would have been great to have had him join us in Australia, but it was the wrong time in his career. He was finishing his degree, he had recently found a viable postdoctoral position, and he needed to write up his doctoral research for publication.

Nancy met with Jean in her university office to discuss her lack of interest and participation in the preparations for Australia, then immediately called me at home to let me know how it had gone. She said Jean had apologized for her lack of availability and participation, but that she was "stressed out" because of trying to prove herself on a highly unsuccessful badger project, and by the competing social demands of Eddie's family. Furthermore, her new boyfriend was taking lots of time. "It will be great to get away from all of this," Jean said. "I can't put much time into preparations now, but when I get to Australia I'll give 100 percent effort. You can count on me."

In early November, Nancy had a call from the chair of the Department of Ecology and Evolutionary Biology at the University of California at Irvine. He said that among some ten candidates that the department had interviewed for a position, she had come out on top. Her name had then been entered into a university-wide competition for the select position. Nancy had again fared as well as she had in round one. What, the chair asked, would it take to entice her to leave the corn and soybean fields of Illinois and come west as a tenured full professor?

Nancy had visited the Irvine campus only once, more than a year before. She had been invited for a seminar, to see if she was interested in the department and vice versa. She had been there long enough to be impressed by the research credentials of some of the faculty, by its youthfulness, and its recent growth. But like scores of others from the Midwest and the East and the South who were interviewed by California universities, upon her return home Nancy expressed shock at the exorbitant cost of housing in southern California. We joked about the prospect of moving to a city that had only been invented in 1971 and was famous in high school history books as far away as Kzyl-Orda in Mongolia for being America's most completely planned city.

With a job offer from Irvine in hand, we remembered that at Illinois Nancy had a huge new lab with long indoor flyways and wide-open breeding spaces for her finches. Since depressions had become an accepted fact in American universities and Nancy was not a gene-tech genius who brought in million-dollar grants, it seemed unlikely that any other university would come up with a quarter of a million dollars to match her space-rich birdland. But as so often happens when a Napa Valley chardonnay begins to take effect, we got young at heart and said, Who knows? And I said, "Wouldn't Lotusland be great for Cole and for our insatiable appetite for exotic foods and fine wines?"

I found myself thinking out loud about why a Mediterranean clime is called Mediterranean, and how southern California would certainly be better for the creaky aches that Nancy increasingly felt in her elbows and knees when the Midwest went into predictable deep freeze. Other sobering facts crowded my brain. For as long as we had been in Illinois we had suffered under a governor who thought that higher education was a program that promoted heavy drug use. In recent years Nancy had been burdened with an unusually high administrative load in her department. Several of her colleagues had given up on serious research, and the better ones had all left for career changes or more promising academic opportunities elsewhere.

Nancy went to see the director of the School of Life Sciences to inform him that she had a firm verbal job offer. The director wanted to hear about it. He was especially eager to hear the salary she'd been offered. She told him the amount, which was significantly greater than what she was currently making. The director said that he was anxious to make a counteroffer and that he would give her $10,000 in salary more than the University of California was proposing. In addition, he wanted to see her wish list as soon as possible. Would she like a research assistant? Some new lab equipment? Extra salary for a couple of summers? What?

The director said that his generosity was not merely a reflection of the school's desire to keep Nancy, significant as her research and administrative contributions had been since she arrived a dozen years earlier. Rather, the school and the university, he could now candidly admit to her for the very first time, "has taken advantage of you." She had not been justly compensated for her contributions. Relative to her male colleagues and others in the school, she was obviously underpaid. "The time has arrived to rectify past injustices," he stated.

Before Nancy left the director's office, he said that he would need to see the written offer on University of California letterhead. Without it, the salary increase and other perks would not be forthcoming. After all these many years, perhaps Nancy had finally decided to heed the well-worn truth that the only way any professor ever gets a decent salary raise at an American university is to march into an administrator's office with an offer from a competitor.

I was furious when Nancy told me about her conversation with the director.

Two days later, with numerous details of the University of California offer still to be worked out, I put our house up for sale. Three days after Nancy had talked to the director I was packing boxes and planning the move.

2

The Search for Breeding Redbeaks

I had two months in which to find at least one sizable breeding colony of zebra finches. By my departure date, Nancy had managed to get leads to six promising sites in the Northern Territory. But how reliable were her sources? In late February she had called one of Australia's top ecologists in Alice Springs, who worked for the Commonwealth Scientific and Industrial Research Organization, the country's premier government research institute. He was familiar with the ecology and mating behavior of zebra finches. He had said, "It looks real good here. It's warm and the grasses are seeding. The finches *must* be breeding." Two days later he called Nancy, embarrassed. He had checked further and found no evidence of breeding.

On my way out the door, Nancy said, "You'll find them. Just imagine you're going muskie fishing."

I laughed. The last time we went muskie fishing was in the first days of October 1988, on Wisconsin's Chippewa Flowage, a world-renowned muskie lake. Nancy spent our nine days there in a warm lakeshore cabin analyzing finch data. I, as possessed as the craziest of muskie fisherman, spent eight to ten hours a day every day drowning suckers thirty-five centimeters long and throwing twenty-centimeter lures at weed and reed patches and sandy shorelines. Several days I had ice on my pole and line, and in my beard. By late afternoon on two of the days, I was edging perilously close to hypothermia. I didn't get a single hook into a legal-sized muskie that trip.

In Cairns three customs agents in short-sleeved blue shirts, tan shorts, and cream-colored, knee-high socks encircled the pine crates and lowered their eyes. The fourth agent sidled over to me and said, "What do you do?"

"I'm a biologist."

He said that he had recently taken part in netting chickens north of the city that were suspected of harboring Newcastle disease. He explained how the netting was done. I sensed that he was searching my face for an intelligent response, proof that I was in fact a biologist. I showed interest in his story. But I knew virtually nothing about Newcastle disease.

One of the other agents approached me—a bulky, middle-aged sort with a grisly brown beard and a floppy beer belly. He got down on his knees in front of one of the crates, then took a knife out of his pocket and burrowed small holes in the soft pine.

"They're bird traps; I'm going to be trapping zebra finches," I said to him. He ignored me and continued to mine the wood for heroin or cocaine, or—more likely—an unwelcome foreign insect that could wreak unforeseen havoc on native plants and animals (as have so many exotics introduced into Australia).

I volunteered a description of what the traps were for and how they worked. Halfway through my explanation, the man stood and shrugged his shoulders. The chicken netter took the cue and said, "Okay, you can go."

I loaded the six walk-in traps and my huge overstuffed bags onto a steel cart and wheeled away. Customs agents hadn't so much as poked a pen between the trap panels to see if they were indeed what I claimed. They hadn't touched the zippers on my suitcases or small backpacks, which were full of banding pliers, colored leg bands, and hundreds of tiny vials for storing blood. I never could remember the name of the chemical in the vials, a twenty-letter something or other that would preserve the blood until it was returned to Nancy's lab to be run through a DNA fingerprinting protocol.

I exchanged four hundred dollars in greenbacks for Aussie money—lavender five-dollar bills, apple-green tens, mustard-yellow fifties—and then flagged a taxi. "Take me to the Barbary Coast," I said to the driver. "I'd like an old hotel, something with character."

I got a corner room that fronted on the balcony, next to a cripple who, I'd soon discover, lived on welfare and cheap whiskey and bananas. My room was clean, the sheets pressed. The bed had broken springs, a saggy mattress. A tiny metal sink had a mirror big enough to allow me to shave. Above the bed was a clunky fan. The toilets and showers were down the hall and around two corners. I dragged the bird traps up to the balcony and parked them outside my door, then headed for the street.

I found several trees that had been taken over by lorikeets, who chattered relentlessly. The noise they made was an order of magnitude more penetrating than a park full of cicadas on a steamy Midwestern night. They came into the trees in swarms of twenty, thirty, fifty. They circled, they dove, they landed. They filled the graying sky. I lay on my back for the better part of an hour, my binoculars trained on their movements and on their brilliant yellows and oranges, purples and blues. A stick-legged Aboriginal woman walked by, within inches of my face. She stopped and stared down at me as if I was crazy.

"Do you know the names of these trees?" I said, pointing to those taken over by the lorikeets.

"I don't have a clue, boss. I'm not a tourist. I'm a local."

In the morning I walked over to automobile row and went into the first lot I came to. When a salesman approached me, I said, "I've got $A10,000 in my pocket to buy a used station wagon or van. I need something reliable, with decent clearance. Something that can take a beating on several thousand miles of unpaved station roads."

He showed me three or four cars and a couple of Landrovers. Each time the price was well above $A10,000. Each time I reminded him of what I wanted and how much I had to spend.

He showed me an eight-year-old Holden station wagon with a rack on top. It looked roomy, about what I had in mind. I wasn't sure it had enough clearance for some of the roads I imagined I'd be on. When he started the engine, it sounded tinny and untuned—like trouble. "Only needs a tuneup," he said. "I'll take care of that straightaway. Come down to my office; it'll only take a half hour or so."

He sat back in his office chair and pulled on the graying hairs that stuck from his open white shirt. He began his sad tale. He said that until recently he had devoted all his time to his investment properties, of which he had several in Cairns. He had imagined that within a couple of years he'd be able to sell his properties to Japanese investors and take early retirement on Queensland's sunny Gold Coast. But disaster struck. Airline pilots went on strike and stayed out for eighteen months. They were eventually all fired. By then tourism, the economic linchpin of Cairns and much of Queensland's north coast, had gone into a deep slump. Investment properties lost half or more of their value. Many were vacant and no one wanted to buy them. The mortgages on them were more than the present market value. Now this former fat-cat speculator and investor turned used-car salesman was thinking of filing for bankruptcy. To make matters worse, his wife had just left him. His concerns of the moment were paying rent and food bills and delaying the government's demand for back taxes and the bank's for principal and interest payments.

An hour had passed since the Holden went to the garage. Somehow I couldn't find the energy to get up out of the chair. My lethargy got me more stories: about how much of Cairns the Japanese owned, of how they went through five or six hardheaded negotiations before finally striking deals that gave them the best possible terms. He didn't blame the Japanese for his woes, but he left no doubt that he felt he was a hostage to Asian money, that the invasion had just begun.

Fascinated with all the lorikeets I'd seen the night before, when they'd filled the sky and trees in back of the hotel just before sundown, I shifted gears. "Do you know anything about the screeching rainbow lorikeets perched in trees all over town?"

"Yes, I know all about them. There used to be two trees out front and you could tell when they were dying. The more lorikeets you saw, the more you knew those trees had had it."

Biology at its best.

I called an hour later and the car still hadn't been tuned. I decided to look elsewhere.

I walked several blocks west and wandered into a flashy, flag-covered lot full of new cars, old cars, vans, four-wheel drives. Forty minutes later I counted out more than $A8,000 in cash and signed some papers and insurance forms, for an ugly green 1983 two-wheel-drive Toyota van. I bought it because it didn't have a low wheelbase; it would carry several bodies, lots of equipment. It looked resilient, as if it would take abuse.

I had wanted to drive away in the van immediately after signing the papers. But the middle-aged salesman feared that a few minor scratches on the windshield might impair my vision. He said he didn't want to be responsible for the death of a Yank who had absolutely no idea how hard and dangerous it would be driving thousands of miles on dusty and corrugated sun-drenched station roads. I was further delayed when the shop mechanic had trouble finding a replacement for the rear windshield wiper.

I had supplies to buy: some for myself, some to help find the birds, most for the imagined fieldwork. With the help of the telephone book, bar and street phones, the road, I located surplus and hardware stores. I bought a two-meter aluminum ladder with extension, a small collapsible table and five folding chairs, a camping stove, pots and pans, jerricans, and water containers. From a garrulous woman who smelled like Nancy after a long day's work with zebra finches, I bought six kilos of finch feed, four chicken waterers, five bird feeders, and seven roomy double-compartment boxes for holding trapped finches.

I spent several hours driving around Cairns looking for a trailer in which to haul everything that wouldn't fit into the van once Nancy and Cole and the team arrived. I finally decided against the purchase, on the grounds that the only one big enough to call a trailer—about the size of a miniature U-Haul—cost nearly $A2,000. Besides, its size and shape would make it hard to negotiate bumpy station roads.

With a couple of hours before stores closed, I went looking for topographic maps with enough detail to get me to a station house or a creek or a bore with water, should the van go silent and refuse to be resuscitated. I found none, even in a shop that sold nothing but maps of every conceivable size, color, and subject matter. Nor did I find anything worth buying after carefully examining state-of-the-art maps of where zebra finches had been captured in Queensland, the Northern Territory, and Western Australia.

When the afternoon tropical downpour began, I stepped into a bar. Soon I was listening with great interest to a believable story about a taipan that chased a farmer across his veranda, then killed the farmer's dog when he came to his master's rescue.

The downpour suddenly stopped, and I wandered over to a camping-gear store. There I heard another taipan story, this one about a tourist north of Cairns who chased the highly venomous and aggressive elapid, which turned on him and bit him a couple of times. The man was dead before he got to a hospital.

I wondered if Nat would heed my words if I told her not to get too close to snakes. Later I reminded myself that my mind was probably working overtime for naught. It was highly unlikely that we'd trap finches where taipans were found. We *would* be well within the range of king browns. Still, despite their reputation for boldness, they're seldom seen. Nancy had spotted only one in her previous year-and-a-half field stint; I'd seen none on that stay.

A more likely problem was contracting a rare tropical disease. A recent issue of *The Bulletin,* Australia's version of *Newsweek,* had an article with the biblical title "When the Floods Came." It was all bad news. Much of the northern half of the Northern Territory had just had its wettest Wet in more than thirty-five years, some two hundred ten centimeters in Darwin in the previous five months. Vast valleys were flooded as far as the eye could see from a helicopter. The standing water was a breeder of disease and a welcome haven to man-eating crocodiles that had escaped from rivers and farms. Already nine people had died from a bacterial disorder known as melioidosis. Twenty-five cases of the rare tropical disease had been identified; there had only been ninety cases in the previous thirty years. Allegedly, outbreaks of gastroenteritis and hundreds of confirmed cases of Ross River fever were rampant throughout the Territory. I couldn't be certain whether or not a sprawling section of the Territory along the Western Australian border where we had once trapped finches was within the area described.

Uneasy about driving on the wrong side of the road, I drove the first eighty kilometers south of Cairns as if I was afraid of being stopped for drunk driving. Among the dense and lush cane fields, the weedy second growth, and the swollen streams, I searched for birds large and small, certain that I'd be able to name or recognize no more than one or two. The vastness of my ornithological ignorance is exceeded only by my thirst for surprise, my lust for novelty.

In Ingham, a lovely little country town with wide streets, no Japanese or American tourists, and a complete absence of signs advertising discount opals, I got a hotel room with thick wooden double doors that opened onto a veranda overlooking a tree-lined median strip. After working on my journal—screaming lorikeets in the trees outside jogged my memory—I went down to the wraparound bar and had an undefined grilled "reef fish." My belly full, I made my way around the bar, asking anyone whose attention I could get if they were aware of zebra finches anywhere in the area. No one seemed to know whether a zebra finch was a bird or a four-legged furry animal. I gave up and settled into a quiet beer.

I crossed the street to another bar, where I plied the same bird question. Same answers, more or less. And more than I'd reckoned for. A barefoot, twenty-three-year-old woman with eight or nine earrings in her ears offered some unsolicited demographic opinions on the town. She said, "What we mostly got here is Italians, Spanish, and coons." I raised my eyebrows and she said, "Coons, luv. Niggers, I guess you call them."

I called Nancy, let her know that the trip across the Pacific had been easy, that the hardest part had been the shuttle-bus driver who wouldn't allow me to board with the bird traps because they were too big and too clumsy. He changed his mind only when I stuck a ten-dollar bill in his hand.

"Cole has another ear infection," she said. "His temperature is 102 degrees." She was worried. This was his third or fourth infection in the last couple of months. She didn't put much stock in the doctor's telling her not to worry.

I picked up a newspaper and read that two hundred fifty thousand Australians in a population of 17 million had cellular telephones. A little farther on I discovered that according to one fastidious letter writer, by being in Queensland I was not (contrary to a long-held belief) yet in Australia's Top End. He pointed out that the nation's bible on such matters, the *Macquarie Dictionary,* said that the Top End referred to the northern part of the Northern Territory. It extended only as far south as Katherine.

The day was hot and dry, and when in Townsville I got out to quench my thirst I was attacked by aggressive flies—an undeniable reminder that I was back in Australia.

By early afternoon I found the offices of the Department of Primary Industry in Charters Towers. One person led me to another, that one to another, until I got to the desk of David Mims. After I told him what I was after, he brought out some topographic maps and moved his fingers and my eyes to several sprawling sheep stations within a hundred kilometers of Winton, Queensland. He knew these stations intimately, from his teenage days in the 1980s, he said. Thousands of zebra finches lived on these stations; they were everywhere.

I took lots of notes, bought maps and marked them up. I underlined the word *mimosa,* a common name for a prickly acacia loved by zebra finches for the protection it affords against predators. Mimosa is common around bores and artificial water, as well as badly disturbed land on cattle stations.

After a long drive through low open woodland, I stopped at the Grand Hotel in Hughenden for the night. The woman in the bar who rented rooms wanted to give me number 13. I said, "No thanks. With what I'm after, I don't need any bad luck."

"What's that you're after?" she asked.

"Zebra finches." I described them.

"Oh, redbeaks!" she said. "They're everywhere around here. Hundreds, thousands. *Every*where!"

She gave me room number 12. Another room that fronted on a veranda, this one overlooking a naked dusty Broadway, not a human or a car in sight.

I showered, then headed downstairs to find out exactly where I could find Shangri-la—a thousand sex-hungry breeding redbeaks. No one had specifics. Everyone agreed that all I had to do was take any road leading out of town and I'd be sure to bump into just what I was looking for.

I did, and found nothing.

Before turning south at Hughenden for the ninety-kilometer trip to Winton via Oondooroo, I stopped at a tire shop and bought two meter-long pieces of rusty metal sheeting to throw under the van's tires. Without four-wheel drive or a winch, these would have to be my answer to getting out of sand or loose dirt, were I to get stuck in some isolated place.

For the first fifteen kilometers, the road was lined with flowering prickly acacia with an understory of buffel grass. The bright yellow flowers put me in a funk. The trees and the alien grass, a real imperial sort, were signs that this land had been badly abused by livestock grazing.

Once beyond the thickets of yellow acacia, the road became single lane, a deeply rutted washboard that had me stopping constantly to walk hinter and yon to some promising bush or tree in search of finch nests. After a long morning and afternoon of sticking my arms through tunnels of sharp spines, I hadn't come upon enough nests to fill a small lunch bag. The biggest catch of the day was one I didn't want to tell anyone about; all the birds I'd been killing with my rolling wheels. My noisy approach flushed them from the tall grass and weeds that lined both sides of the road.

I consoled myself by taking note of those I saw through binoculars or with the naked eye: peaceful doves, diamond doves, little corellas, pallid cuckoos, rainbow bee-eaters, willie-wagtails, brown falcons, gallahs, a magnificent bombing squadron of budgies, and a half-dozen black-crowned bustards huddled in tall grass.

I invited myself into Vellum Downs Station, and soon I was talking about zebra finches with a friendly long-faced woman in calf-high black rubber boots. Yes, they came now and again, and sometimes in great numbers, she said. But she hadn't seen any lately and didn't know where to find them when they were breeding. What she knew a lot more about, and wanted to talk about, was king brown snakes. Recently her teenage daughter had seen eleven of them in one day on the front lawn. The previous day she herself had killed a big king brown that was crawling up her nine-species aviary—the same snake, she was sure, that had killed two of her cockatiels a week earlier.

In Cornfield I stopped to find the owner of the blood-drenched four-wheel-drive Toyota in the parking lot. I wanted to see if he'd tell me how many kangaroos he'd shot the previous night.

The two shooters were the only people in the unlit bar. Both were covered with blood. John, with the loud voice and the proud swagger, wasn't a day over twenty-five. His uncle, leathery and spotted like a dalmatian, claimed to be eighty-eight. He owned a nearby station on which the two of them had shot and skinned fourteen red kangaroos during the night. For their efforts, they'd get ten dollars apiece for the hides and nineteen dollars each for the carcasses from exporters and pet-food vendors.

John said that they worked sunup to sundown, and on a lucky night they'd kill and skin forty or more kangaroos: grays along streams, reds out in the open, head shots only. After they downed the first six or eight, they'd pick them up and skin them, then begin again. I quickly discovered that their principal preoccupations had nothing to do with clean shots, quick kills, chasing down wounded kangaroos, and maintaining populations that could be exploited indefinitely. All that seemed to matter was whether or not they were getting easy shots, using the right bullets, and firing as few as possible. They were unhappy that the bullets were costing them twenty-five cents apiece. Life was hard.

Oondooroo, established by a Dutch family in 1883, was once a sheep station that stretched from Winton almost to Hughenden, two hundred kilometers to the north. At one time, more than a hundred thousand sheep roamed the station lands. Now the number was around fifteen thousand, and decreasing. In two years a bale of wool had dropped from $A1500 to $A200. Nor was there a decent market for mutton and lamb. The most anyone was getting from the abattoir in Winton was $A14 a head. The economy was bad, real bad, and the "bloody politicians" were to blame. Now at a modest twelve thousand hectares, the station was up for auction. All that was needed for a change of ownership was a bid of $A15 a hectare.

After an exchange of seat-of-the-pants opinions on marijuana market prices in the United States and Australia, I inquired as to the whereabouts of flocks of breeding zebra finches. Yes, there were finches here and there on the station, and every so often kids from Winton came around with their guns asking if they could shoot them—and anything else that moved. As for zebra finches, they once allegedly numbered in the thousands; they would fill the sky. You could find them everywhere on the station. But now you didn't often see many. Some here, some there . . . go ride around and see for yourself.

I did. I drove one lumpy station road and then another, dropping markers, making sure to remember trees or bores so I wouldn't get lost. I found a handful of old nests. Used finch nests survive more or less intact for a year or more, cemented together by the nestlings' droppings. The old platforms are dark colored and relatively easy to spot. Finches prefer to build new nests on the old platforms. There was no sign of recent activity.

In Winton an inquiry about redbeaks elicited the information that I'd better start referring to them as "zebras" (rhymes with "Debras") or waxbills. That was how they were known locally. It soon became apparent that there was as much confusion about what locals called zebra finches as there was about their abundance or their location. One tale was that there hadn't been many around since a bad drought in the 1960s. Another said that waxbills were everywhere, but that I'd come at the wrong time of year; this wasn't the breeding season. Which occurred in . . . September? December? January? No one was quite sure. And then through an old swagman, a gaunt man of uncertain age who resembled Rip Van Winkle from his nose down, I was given the name of a Barry Cooper who knew where to find "millions of finches."

I found the man at a bustling bar on Winton's main street. Cooper quickly confirmed that he most certainly did know where to find millions of finches; they were at the airport. He drove me out to a bore near a runway to show me the millions that were there at any time of the year, he said. We stayed for an hour and walked all over and saw less than a dozen. I found fewer than half a dozen old nests.

On the trip back to Winton, Barry informed me that if zebra finches didn't get water every hour they died. He said that Aborigines followed the birds to find water. The Aboriginal part of his knowledge I knew to be true, from reading old ethnographies. That zebra finches needed water every hour was a fact that would surely rattle Nancy's sense of reality.

I spent a couple of days scouring the countryside around Winton in search of zebra finches: redbeaks, blackbeaks, their nests, flocks of them. I drove through sheep country, all the way to Longreach. I headed south down the awful Jundah road, through Wamambool Downs, past the Kangaroo Mountains. And then I crawled up the Landsborough Highway, on toward Cloncurry.

Twenty-five kilometers north of McKinley I found the only thing that remotely resembled a finch gold mine. On the west side of the highway near a large blue billboard announcing "Historic Cloncurry" were large and small clusters of prickly acacia. The largest cluster, some thirty-five to forty trees, encircled a livestock watering hole. More than 90 percent of some three dozen nests were in this cluster. One large tree had fourteen; I reached several of them. Only two suggested activity within the last month to six weeks. I couldn't determine the ages of the other nests, but they were old. Within the one tree with most of the nests, they were clustered, more than two meters above the ground, toward the center. Smaller trees, and those away from the main watering hole, had few or no nests. Avoiding predators—snakes, lizards, birds of prey—was high on the zebra finch list of priorities. Natural selection was working its usual wonders.

I found a few redbeaks, but no blackbeaks. No reproduction had occurred recently.

West of Winton, on the long looping road that would take me to Mount Isa, I drove through a landscape populated with magnificent gidya trees. Great to see, but dreadful finch habitat, not even worth stopping and bringing binoculars to my eyes. Then for kilometers on the one-lane road I encountered neither trees nor cars, nothing but a meter-long goanna in no hurry to get out of my way.

My mind got lost in the distant tan and brown and shadowy hills, the red, the brown, the black mesas. These were memorable landscapes, the stuff of dreams, too beautiful for postcards. I felt groggy and I needed a break from the road—something to eat, a nap in a patch of sun-drenched grass.

Trees and an unusually large building appeared out of nowhere. I pulled up to the Middleton Hotel, ordered a hamburger and was quickly introduced to six raucous, hairy-legged men who worked for Telecom Australia. They were in faraway Middleton to put in a soaring telephone tower at a cost of half a million dollars. This little construction would make it possible for a grand total of ten families to consign their radio phones to the closets of history.

That day I learned that tiny Middleton harbored someone by the name of Peter Naylor. Peter owned seventy guns and shot anything that walked or moved or breathed: foxes, dingoes, pigs, 'roos, and feral cats. Cats around Middleton were big. I too would want to shoot them on sight, because they killed any bird they could get near. It was also worth shooting them because each skin was worth seven dollars. And that, mate, is bloody fine grog money.

I learned that just east of Middleton, seven thousand sheep had recently been shot and bulldozed out of sight. Who knew how many tens of thousands were being shot elsewhere in Queensland, New South Wales, South Australia? The market was glutted. Reality ruled: it was cheaper to kill the sheep and pay the driver of the bulldozer two dollars a head to push them out of sight than to raise them.

Kevin, the good-as-they-come story-telling publican at the Middleton Hotel, was eager to show me the local countryside and help me find colonies of breeding zebra finches. "I'll show you plenty of redbeaks," he said. "They're everywhere."

We stopped at a bore, we followed a creek bed, and we crossed and recrossed lines and clumps of coolibah trees, bean trees, and prickly mimosa, desirable finch habitat. There was lots to see, to remind me that I was very much on a continent unto itself. Six emus, eleven gray kangaroos, seven bush turkeys, plovers, scores of galahs, lots of flies all over us when we got out and walked. We saw exactly five zebra finches, about two orders of magnitude less than the number Kevin had told me we'd see. But the evening was young and I shouldn't despair yet, Kevin said.

We bumped and rolled over the high grassy downs, then slowed for a close look when we came upon nine freshly skinned kangaroo carcasses. More than good enough to eat if we'd been hungry and lost. "And a lot better than oily emu meat, too." Kevin explained that there was no market for kangaroo meat. A good 'roo shooter could make a hundred a night, he claimed. The government didn't want just anyone shooting and skinning kangaroos. You had to kill two hundred a year to keep your license.

We swayed and jerked along a sunken creek bed, still looking for the promised finch nests. Kevin looked, I looked. We stopped and got out and walked around. No old nests, no new nests, no nests at all. Not even in the usually promising mimosa around the cattle troughs.

We meandered through uncluttered gidya, between a pair of long-leafed white woods, among gray coolabahs. Galahs flew overhead, little corellas screeched, rosellas and hawks perched on distant branches. The tires crunched the black bed of sharp pebbles. Then we whirled around and shot down into yet another dry creek bed. Two gray kangaroos—a mother and a joey—darted toward the woods off to my right and disappeared before I could get the binoculars to my eyes.

We saw everything but zebra finches.

The hunt for breeding colonies of redbeaks resumed as I drove southwest, the distant and tawny Finucane Range off on my left, the Diamantina River of legend and outback song way off and out of sight to my right. In the vicinity of the Hamilton Hotel, the first significant human sighting since Middleton, my legs and eyes found a total of three male zebra finches flitting about high in a gigantic pine tree.

Boulia, hot and desolate and dead, was a decision point. Which way to go? I made inquiries in the Australian Hotel and Motel and learned enough to convince myself that heading south to Bedourie and vicinity in search of finches would be a mistake, and probably a waste of time. The road was poor, and the Georgina River, which I'd have to cross, was rising. I decided to play it safe and continue on to Mount Isa.

I crawled over another 120 kilometers of highway, making frequent stops, using binoculars, going over or under fences, through grass short and tall, sticking my arms in all kinds of trees in search of eggs or signs of recent nesting. I got two fingers inside one finch nest, and it was old. How old I didn't care to guess.

The fruitless search for breeding redbeaks went on. North and south of Camooweel. A few here, a few there. Mostly nothing.

Five kilometers east of the James River I slowed to look at a large red kangaroo sitting on the left side of the road. Sensing that something was wrong, I drove right up to him and got out and approached him. He had a broken leg, perhaps a shattered hip. He'll cook in the hot sun and die of thirst and starvation, I thought—a painful death. I looked in the van for a weapon, something big enough to kill him. Everything I had would have made a mess, and I might have botched the job. I went back to my road atlas and saw that luck was with me; just west of the James River was a square red dot that read "Avon Downs Police Station."

Three cops were sitting at small square desks, relaxed, laughing. When I told them about the problem, they said they'd take care of the matter right away. I assumed they'd shoot him.

I didn't want to see the kangaroo sitting there all day; my pessimism genes overrode those telling me to trust what I'd just heard. So I drove back to the nearby river, parked, and headed for the bridge. I took out binoculars and watched a hundred or so little corellas screech while black kites rode vast thermals above a river clotted green with water lilies.

When the policemen returned, they said they'd put him down. I felt relieved.

Less than two hours later, I again pulled over to the side of the road when I saw a gray-haired Aboriginal man and his puffy-faced female companion standing beside a jerrican, waving. They wanted enough gas to get from the nearby Aboriginal settlement, Arruwurru—barely visible on the horizon to the south—to Camooweal, and the bank and the grocery store. The old man, whose deep, sad eyes seemed to be floating in an equally dark pond, said they hadn't had any real food for a week. I believed him. I emptied my jerrican into his.

In Tennant Creek, in the hotel of the same name, I met a bright-eyed woman by the name of Jessamine who, in addition to tending bar, wrote for the local paper, the *Tennant Times*. It quickly became obvious that she had the pulse of matters Aboriginal and local, and what she told me sounded reasonable. It all fit with what I'd learned in my stay in the Territory in 1986 and 1987, and what I'd repeatedly come across on this present visit.

Jessamine said that Aborigines were becoming increasingly militant in their dealings with Europeans. That was, she opined, to be expected. Their heralded and controversial land rights, formalized in legislation in the 1970s, were now secure. The Territory's Aborigines had land, lots of it—and though marginal in an economic sense, land it was. They also had plenty of people to take up their case, to fight large and small wars on their behalf: academic anthropologists, self-styled anthropologists without credentials, social workers of a hundred persuasions, and, not least, those romantics and idealists keen on fossilizing for all time an Aboriginal society they could only imagine because it probably never existed. The aim of many of these outsiders—from Sydney, from Melbourne, from the farthest reaches of Australia—was to educate Aborigines about the pervasive racism, the history, and the continuing colonialist aims of their oppressors, and to show them how they should fight all these injustices. And so, properly sensitized and inculcated with abstract notions of oppression and a lot more, Aborigines were becoming more vocal and more violent. But also more confused.

Aboriginal elders and parents were fast losing their children to boom boxes and designer jeans, to Michael Jackson and video games, to ice cream and the gratuitous violence found on imported American videos. The Aboriginal Dreamtime, their oral history which is chockablock with totemic ancestors and sacred places and a whole lot more, was almost as foreign to the newest generation of Aborigines as it was to vacationing Americans.

In places like Tennant Creek it was easy to see Aborigines as unrecoverable alcoholics. Their camps in the surrounding countryside were dry. So they drank in town, in public—mostly on the street, on the medians, in the parks. And when they drank heavily, they loved to brawl and break windows, and sometimes take what was behind the windows. They added to their woeful public image by defecating in public every so often. Those running Tennant Creek responded to all of this by employing roughly one policeman for every hundred people in the population.

Whatever apartheid I saw in living quarters, I was told, I shouldn't conclude that whites hadn't tried to live alongside Aboriginal peoples. They had, sort of. But every time an Aboriginal family moved into a white neighborhood, the one new family soon became two or three or five. (This, after all, is the nature of complex Aboriginal kinship systems.) The numbers of young and old and dogs, and the garbage in the front yards, were bad enough in the eyes of Europeans. Add in the raucous parties, and latent racism quickly became manifest.

My eyes only confirmed Jessamine's dismal observations and journalistic anthropology. Tennant Creek's main street was thick with aggressive and drunk and begging barefoot Aborigines. A disproportionate number were women, infamous here as elsewhere in the Outback for their pugnaciousness. They not only initiated fistfights with men, their husbands included, but won more than their share, it was said.

In one of their town "camps" Aborigines lived outside and around their expensive government-built homes. As elsewhere in the Outback, homes for many were turned into storage sheds, toilets, shells stripped of doors and cabinets, anything else made of wood, because wood is great for building warming fires. Some Aboriginal people have even preferred to return to traditions and ways of life developed long before the coming of the Agricultural Revolution; once more they live beneath crude thatch lean-tos.

All of this, Jessamine said, was happening not just in Tennant Creek and not just in Alice Springs, Central Australia's major population center and chief magnet for faraway tourists, but also in small, easy-to-miss settlements to the north.

A three-hour journey west of the Stuart Highway, out to the Warrego granite mine, was fruitless: no sightings of zebra finches, few of good nesting habitats, nothing to record in my journal from a visit to a bore cancerous with prickly acacia, one of many signs of overgrazing. The Devil's Marbles, an impossible-to-miss stop along the way, was the trip's highlight. The large boulders that appeared out of nowhere have fertility significance to women in the Aboriginal Dreamtime. These boulders are a sacred site, in theory accessible only to Aboriginal women. But, like so much else that was breaking down in Aboriginal society, tribal men came to the boulders: to drink, to play cards, to have male-only powwows.

The hunt for breeding zebra finches—nay, zebra finches of any stripe or behavioral inclination—continued. All leads proved unfruitful; it was becoming easy to conclude that bird knowledge among the masses in outback Australia was all about overheated imaginations and memories of times past that may or may not have existed. It began to seem I would be better off holing up in a hotel with a swimming pool, to begin a never-to-be-finished read of *A la Recherche du temps perdu*. I mused: Should I call Nancy and tell her to cancel the students' trips? No, we could regroup. How about doing something on doves, or parrots, or wagtails? We could, couldn't we, come up with a new fieldwork project?

By the time I arrived in Barrow Creek, I was so tired that I paid no attention when I rented a room for the night that could only be described as looking like a commercial refrigerator on the outside, and on the inside not being big enough to do much more than hang three or four sides of beef. As luck would have it, it was the coldest night since I'd left the Midwest. My thin blanket was not nearly enough, but I couldn't gather the strength to wander into the night in search of my van and my sleeping bag.

Next day in Ti-Tree (a tiny cluster of homes, downscale grocery store, police station, and Texas-style bar), I remembered that five years earlier those who owned the bar had had quite strict rules about when Aborigines could and could not drink. What they didn't have were rules on tribal peoples who wished to commit suicide by getting drunk and then sitting in the middle of the Stuart Highway hoping to get run over by anything that came speeding their way. When two of the local policemen tried to prevent this self-sanctioned promotion into the Dreamtime's hereafter by a young Aboriginal male, some fifteen Aboriginal women picked up frozen kangaroo tails and clubs and chased after the cops with the intent of sending them into their own eternal dreamtime. It was only prevented because the local cops had shotguns and showed the women what would happen if they didn't follow orders. The women—locally known to whites by the ancient pejorative *gins*—came to their senses. This had happened a week or so before my arrival, I was told as I nibbled on a chicken sandwich in the bar.

In late 1986, my last time through Ti-Tree, the enveloping station also known as Ti-Tree was being managed for the Aboriginal owners by a couple of American expatriates from Montana, Craig and Sylvia Steen. When I'd chatted with Craig and Sylvia at that time they were early in their tenure, and, while optimistic, they were already having their share of problems. Paramount among them were tribal owners and locals who only worked when they wanted to, or stopped working because in true Aboriginal tradition they had to share everything they made with all their noncapitalist kinsmen. Craig had had to sell off the station's cattle for almost nothing, because they were infected with tuberculosis. To turn things around, he'd have had to rebuild and repair fences, build up new herds, and (not least) develop a work ethic that was as foreign to Aborigines of the late twentieth century as Marxism is to American cornbelt conservatives.

Well, I could go out to Ti-Tree and look around for zebra finches as I had five years earlier, I was told by the publican, but I wouldn't find Craig and Sylvia. They were gone, having left four weeks ago. They'd mended the fences and put in lots of new ones, and they'd reestablished a viable cattle population that numbered around three thousand head. But they'd "had enough."

"Enough" could be summarized by the following story. One day not long before they called it quits, Craig had managed to persuade three Ti-Tree Aborigines to meet him at eight o'clock the following morning to round up cattle. Although there were between twenty and thirty men on the station capable of working, no one wanted to get on a horse or mend a fence unless he was paid $A30 to $A40 more than his reliable welfare check contained. Even then it was hard getting the men to work; they'd have to share everything they made with kinsmen.

At eight the next morning, expecting that the promises to work were legitimate, Craig took several horses out to the agreed meeting site. He waited until eleven. When no one showed up, he and two trusty dogs mustered the cattle. Late that afternoon Craig and the dogs pushed and cajoled four hundred fifty head of cattle toward their final destination. But just short of the yards where the cattle could be branded and dehorned, a couple of the station Aborigines playfully rode through the herd and scattered the lot of them. Disgusted with a pattern too often repeated, Craig howled some choice four-letter words, called his two dogs to his side, and slowly rode home.

Craig and Sylvia's idealism and their patience with people whose values they barely understood just plain petered out. They decided, long after others would have opted for alternative employment, that their life would be easier and they would be happier working at something more or less meaningless—like a roadside bar in Kulgera, on the border between the Northern Territory and South Australia.

I drove straight past the cutoff into downtown Alice Springs and didn't stop until I reached the sprawling grounds of the Conservation Commission of the Northern Territory. Before I got to the front door, I was sidetracked by three dozen red-tailed black cockatoos. They filled a giant ironwood tree that arched over a water trough, no doubt a favorite watering hole. It was hard to pull myself away from the sight of these majestic screamers.

I looked up Randy Jones and Tony Pritchard. Randy was a senior scientist, a low-key, unassuming, but highly knowledgeable biologist who knew the Territory about as well as anyone. I'd worked with him on my previous visit, when I'd studied the vexing problem of what to do with all the feral horses there. Tony, younger than Randy and newer to the biology of the Territory, was the Commission's resident bird expert. Nancy had had some dealings with him in 1986. If he was not as knowledgeable as Randy about the ecology of the Territory, it wasn't for lack of smarts. Certain kinds of smarts are time dependent.

Zebra finches, which are gregarious, huddle together in a laboratory just as they do in nature. Aggregations of zebra finches in nature can number in the thousands.

Nancy quantitatively measuring the hue and chroma of a finch beak by matching it with a Munsell color chip.

The bustard, once common in grassland habitats, has declined dramatically because of grazing sheep and rabbits, predation by feral foxes, and illegal shooting.

One of the many signs of outback isolation.

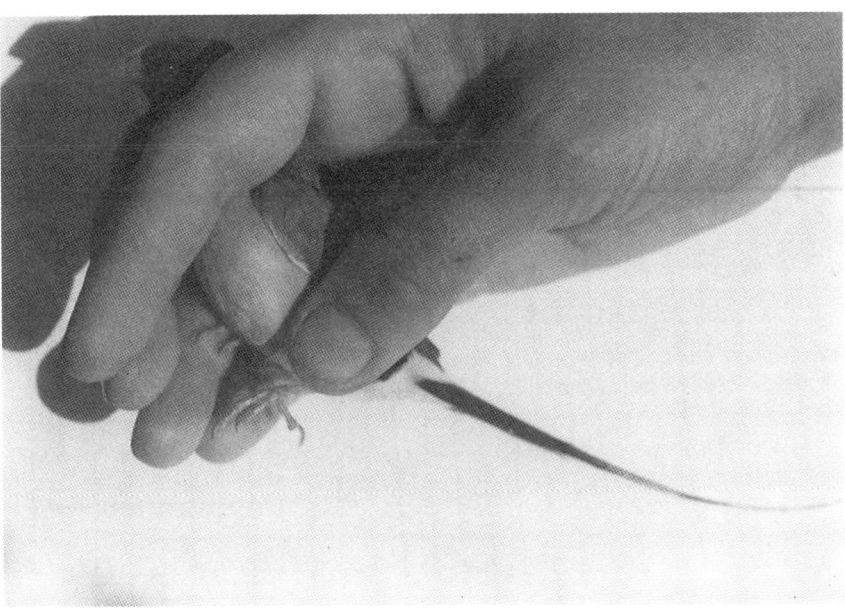

The tails of long-tailed finches vary greatly in length; the tail may even exceed the body length.

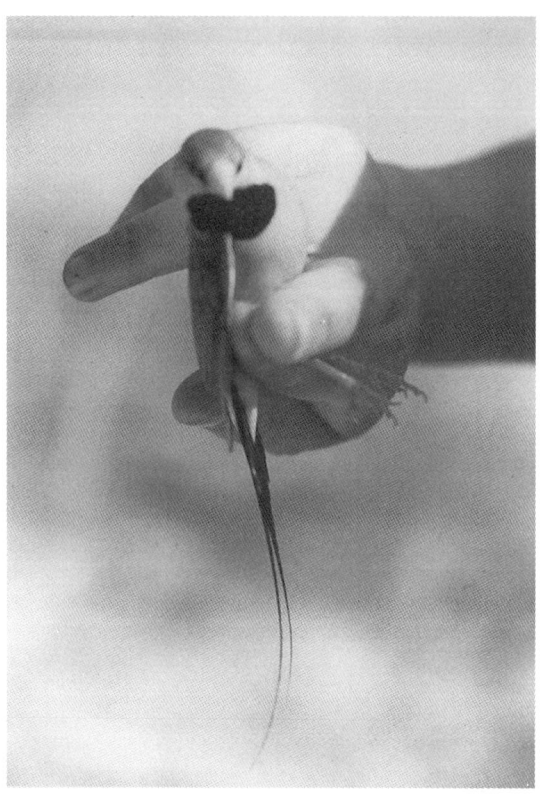

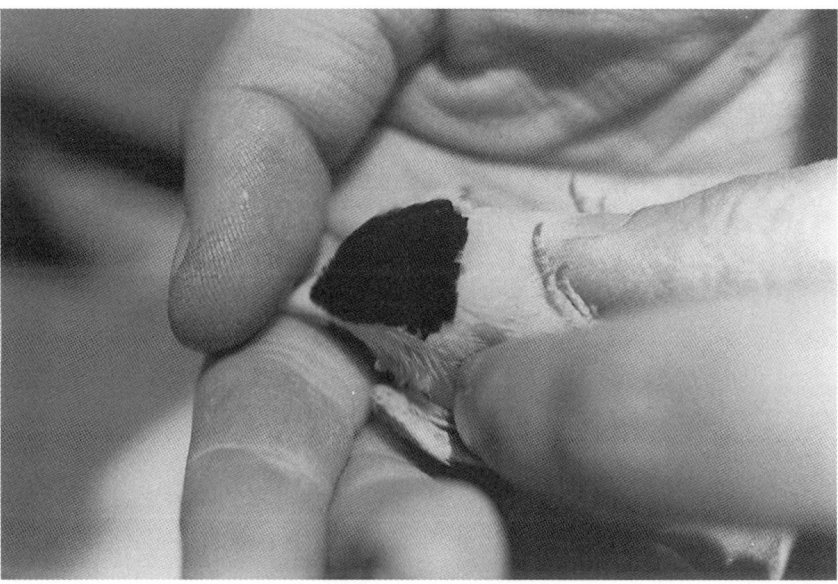

The breast patch on a long-tailed finch; the name *blackheart* may derive as much from the location as from the shape of the breast patch, which is quite variable.

Erosion, caused by overgrazing, is typical of many cattle and sheep stations in outback Australia.

A wedge-tailed eagle feasting on a cow.

The author enjoying a moment of perverse roadside humor after a long and fruitless search for breeding zebra finches.

The Blackheart Hotel, its veranda festooned with hammocks and freshly laundered field clothes.

We talked about zebra finches and about what Nancy and I wanted to do. We discussed rain and lack of rain, and where I might find colonies of finches for the planned fieldwork. Randy and Tony were attentive and thoughtful. But sorry, they couldn't help; they didn't have a clue where I might find what we needed.

My next visit was to the Commonwealth Scientific and Industrial Research Organization. Nancy and I had worked with several scientists at CSIRO and considered a couple of them close friends. I looked them up. Or rather, I walked the quiet corridor and knocked on doors in search of those who weren't in the field or in faraway Canberra or Sydney consulting with government bureaucrats. All I learned was that other than a monthly census of zebra finches done with a couple of walk-in traps on the sprawling CSIRO grounds (a more intense rendition of which Nancy and I and an Australian ornithologist from Melbourne had started up in 1986) there was nothing of note relative to zebra-finch research since our last visit.

I called Nancy and we speculated about finch breeding behavior and what I might do next. Maybe everything wasn't as bad as it seemed? we told each other. I still had the immense area north of Tennant Creek to scout, and that included the Top End of the Territory. Particularly promising were the sites where Nancy and I had trapped with Heidi Blake in or near Keep River National Park. Heidi worked for the Conservation Commission of the Northern Territory. She was now carrying out a long-term study on the endangered Gouldian finch, in some of the very areas with which we were familiar. There were also sites in northern Western Australia where zebra finches were alleged to be abundant. They might be worth examining if I came up with nothing promising elsewhere.

No, we couldn't—we wouldn't—give up. Nancy and I had too much invested, we owed it to the students, we were imaginative enough to come up with a worthwhile zebra-finch project if the one originally conceived wasn't possible. So we agreed: I was to get back on the road, snoop a little harder and farther afield. A bit of luck and these peripatetic chirpers would *surely* be found in the act of reproducing themselves.

Several days of looking at zebra-finch habitats that I was familiar with from our stay in the Alice Springs area in 1986 and 1987 produced nothing of significance. A few nests here, a dozen or so finches there, nothing elsewhere. What had once seemed like a likely last-chance location now looked worse than doubtful sites in Queensland.

I headed north, through heavy mulga country, which is poor finch habitat. After looking for finch habitats in the environs of Daly Waters and finding nothing,

I called Heidi Blake and was told that she'd be arriving at the Conservation Commission office in Katherine the following day. Super news, I thought. Surely, with all the trapping she was doing over much of the Top End, she'd know where to send me.

Heidi was Heidi. Exuberant, sweaty, field weary, glad to see me, all smiles and a big embrace, and the bearer of good news. Very good news. She had just found four eggs in one Gouldian finch nest. Did I have any idea how wonderful this made her feel? I was brought up to date on her project, her research state of mind.

Heidi believes that Gouldian finches are endangered. Once supposedly abundant and widespread in Australia's Top End, Gouldians were now absent from or found sparingly in much of their former range—which included large areas in the northern portions of both Western Australia and the Northern Territory. In the Territory, Heidi's principal concern, Gouldian numbers were not known with any precision, even though she had been trapping them for several years. Her data only showed viable populations in fewer than half a dozen habitats. Her mission: to discover the whereabouts of active Gouldian populations and to find out what needed to be done to maintain or—better yet—increase their numbers.

"Any idea where I can find breeding zebra finches?" I said.

"No. None." She laughed—as if I should have remembered that not for nothing are zebra finches in the scientific literature sometimes described as "nomadic."

If it would be any help, I was welcome to look at her data sheets on trapping finches for the last several years. She had gotten them at various sites, though not, if memory served, in the kinds of numbers that would meet our needs. I could, Heidi said, also get copies of useful topographic maps in Timber Creek, where there was an office of the Conservation Commission and where she and her new husband owned a home.

While talking about Gouldian finches, Heidi had mentioned that long-tailed finches were sympatric with Gouldians, and that she and those working for her had caught many more longtails than Gouldian finches. Without the concrete leads from Heidi that I'd hoped for, I returned to the longtails. I asked what kinds of data she was collecting, what research she had done or was planning to do on them, and whether they were integral to her study of Gouldians. Heidi said that while she and her field assistants banded longtails and recorded their captures, and had a crude system for aging them, she had no interest in them.

I had seen long-tailed finches, with their striking black breast patches, darkly shrouded eyes, and intense red or yellow beaks. Their tan bodies were slim and elegant, made all the more prominent by the very long, wire-thin black tails that gave them their name. Alongside Gouldians, or zebra finches, the longtails struck me as courtly, noble, proud masters of their world. They were anything but overwrought, gaudy, precious—adjectives that came to mind when I looked at Gouldians. But then I hadn't been studying one species for six years, as Heidi had.

The long-tailed finches, I understood as we conversed, got in the way and created extra work. To my surprise, Heidi was content to work on the untested assumption that their behavior, their presence in the very same habitats as Gouldian finches, did not affect Gouldian nesting or breeding activity.

I tactfully pursued the matter of Heidi's lack of interest in the long-tailed finches. A thought forming, I wondered out loud if she'd mind if Nancy and I and our team worked on them—if, of course, I had no success in finding zebra finches.

Sure, why not? No problem at all, she said.

We talked about the obvious, that if we trapped longtails we'd also be getting Gouldian finches in our nets. I assured Heidi that we'd be delighted to help her in any way possible: share data, band Gouldians, administer medicine, work out field problems to mutual benefit. Heidi had several years of banding data on longtails and we could have access to those records, she said. They were raw, she hadn't done anything with them, but we could help ourselves.

Heidi said that I could stay with two of her team members: Neils, whom she described as an intense Dane, and Terry, a tall, lean, eager Western Australian, who had ambitions of one day working as a park ranger. Neils and Terry were living in Keep River National Park, sleeping on cots on a screened porch in a building where guests were received. There they had access to a small kitchen, a pay telephone, a washing machine, and a bathroom. Staying with them, Heidi said, would give me an opportunity to see what kind of data she was collecting on Gouldians, look over the possibilities for research on long-tailed finches, continue my anxious search for zebra finches.

I was elated with the offer. Here, I thought, was a shining example of cooperative science; looked at differently, it was the continuation of a collaborative effort between Heidi and Nancy that had begun several years earlier.

I told Heidi that as auspicious as this possibility was for working with longtails, it wouldn't go anywhere without Nancy's wholehearted interest. As far as I knew, she was still dedicated to long-term research on zebra finches. "On the other hand, we've also got a commitment to four students for work here this year," I said. "We'd hate to have to tell them that the original project busted and we have no alternative, nothing for them to do. It wouldn't be fair to cancel their trips."

The first chance I had, I called Nancy and talked nonstop about the conversation I'd had with Heidi. I barely let her talk, such was my eagerness to sell her on working with longtails. Nancy had researched zebra finches for a dozen years, and though she had an abundance of data that had not yet been analyzed and plenty of years to ask the questions about zebra-finch behavior that were fermenting in her fertile mind, this might be an excellent time to move in a new direction. "You need something different to stay fresh. Why not shift gears now as we move to a new environment, a new job, California." I went on, probably more the hustling used-car salesman than I realized. I argued the case for working on a related, but not too related, species. I talked about the many virtues and payoffs of comparative research. I asserted that the zebra finch was close enough to the long-tailed finch for Nancy to use her exciting findings on zebra finches as hypotheses, as a model for the longtails. "And be young, be Californian, take risks as you always have!" I said, as if trying to convince myself.

"Why not?" she said in her own low-key yet enthusiastic way. "But let me think it through. Give me several days, a week. I have to think about what we could do there in the time we have. And let's not stop trying to find zeebs. We shouldn't give up. I'm still committed to them."

Soon we talked again. Fifty, sixty, eighty dollars worth of telephone time each time I called. Nancy's enthusiasm began to rival mine. "We'd have to do a lot of basic biology," she said. She had done a search in the library on long-tailed finches and found almost nothing: something in German, another small descriptive piece of no consequence. Whatever was known about long-tailed finches was stored in the memory of persons unknown. As with so many species, it was an amateur birdwatcher's landscape cluttered with anecdotes—and the occasional telling, but unrecoverable, insight. There was no reason we couldn't use much of the game plan that Nancy had detailed in her mind, and on paper, for zebra finches, she said. We could still take blood for DNA fingerprinting, and data on age and weight and tail and tarsus length. Besides, Nancy had some ideas about measuring morphological asymmetries, an interest that she hadn't pursued nearly enough with zebra finches. Of course, we would also have all the trap data, behavioral observations, ecological data, whatever we could find out by climbing ladders and sticking our hands in nests.

It was Good Friday, and I was in Darwin. I'd never seen a city of this size so quiet, withdrawn, sleepy. I drove lazily along the bay, walked on the beach beneath the gray sky and out onto a rocky promontory to smell the salty sea-saturated air. The tide was out. I went to an open-air restaurant and found a table that looked onto the street. I drank coffee and more coffee, and read some papers that Heidi had given me on Gouldian finches. Written for popular magazines, with color photos of the Gouldians and of Heidi, they were stories of how she'd become interested in the finches, their endangered status, the medicine she was giving them for an air-sac mite problem that, she opined, was partly responsible for their declining numbers. None of this was science, or even pretended to be. Heidi was courting the public's attention, dancing on that floor where science embraces public support, money of many colors, votes, and semireasoned shouting.

I was beginning to see more clearly just how committed Heidi was to the Gouldian finch. Yes, I thought, she was a less-publicized version of Goodall and her chimps, Fossey and her gorillas, other women who had come upon a sexy animal and made it into a cause, a major part of their life.

I was glad to be on the road to Keep River. Now it was the Victoria Highway, a name whose imagery brought on the urge to read about Victorian morals and colonial manners. I drove hard until I got to Timber Creek, then decided I'd had enough for a while. I'd quickly run down those topographic maps Heidi had talked about, then head over to the pub and wash down yet another day of frustration.

Cold beer in hand, I soon found myself talking with Bluey Janelle, a burly skinhead Aussie wearing swim shorts and thongs and nothing else. He was missing several teeth, and digits on both hands. Bluey was chugalugging beer at Guinness-book speed. He was in the bar with his shoeless three-year-old daughter and his wife. The dwarfish wife was lost somewhere amid the happy guzzling bodies. The daughter was contentedly sitting on her father's shoulders, singing to herself and making funny faces.

A one-time bull chaser, road-train driver, and zoo manager in Darwin, Bluey had come to Timber Creek to pick up some penicillin to treat a three-meter-long saltwater crocodile whose feet were infected and swarming with maggots. The crocodile, one of scores of "problem" crocs that had been caught in Port Darwin to protect swimmers and boaters, had been injured when captured. The Conservation Commission gave the beast to Bluey to help him get his farming venture off the ground.

Bluey was one of a new breed of crocodile farmers—now numbering five in the Territory—whom the government allowed to collect up to 750 viable eggs a year from breeding females in the wild. This permission was given after the applicant forked over $A10,000 for a pro forma proposal stating how the farm would be run and where the eggs would be collected.

Bluey described how he gathered crocodile eggs on the Roper River: with a small boat and a beer cooler, and looking more or less just as I saw him. Except that when he walked barefoot through the mud and swampy breeding grounds he was carrying a .44 magnum with a twenty-centimeter barrel. The female crocodiles were protective. Coming on their nests, you never could tell just how wrathful they might be if you got too close to the forty or fifty eggs they were guarding.

At present Bluey was getting $A40 for each hatchling. This was after feeding them ample quantities of wild donkey and wild horse meat. If he waited two years, he guessed he could sell the growing youngsters for about $A500 each. The Territory was full of these unwanted critters, which he shot and then cut into chinchilla-sized chunks on a bandsaw. Bluey saw lots of promise in the tourist trade. He imagined tens of thousands of Americans and Europeans crowding into restaurants in Darwin and elsewhere in the Territory, all of them desirous of being able to say that they'd journeyed to the land of Crocodile Dundee and had a man-eater for dinner.

Other than the crocodile that needed medicine right away, Bluey was concerned with what to do about the small matter of replacing $A8,000 in equipment that had been washed away a couple of days earlier in a storm. And there was the not-so-small matter of the tent he and his family had been living in; it was sitting in ten centimeters of water.

Bluey and his daughter and his petite barefoot wife left. I bought some supplies and got back on the road. Before I crossed the border into Western Australia I saw four sulfur-crested cockatoos, two dingoes, and more than a hundred hungry red-tailed black cockatoos in mobs of twelve to fifteen. They had congregated around newly burned, still-smoking patches of ground. I counted more than thirty of these charred plots on the long, often one-lane road west.

I called in at the Western Australian Land Management office in Kununurra, hoping that someone there would know where I could find breeding zebra finches in the region. One ranger led me to another ranger, and then to a so-called regional ecologist, who said he didn't have any real information on zebra finches but would make some calls for me. Could I return in an hour or so? I told him I could, but I'd prefer to snoop in his library for material on finches, if that was okay. He nodded and showed me the way to one of those kinds of places that I think of as heaven on earth. When I found nothing on finches, I turned to crocodiles. I was hungry for more information on what someone like Bluey was up to.

Two species of crocodiles are found in the Territory: a saltwater variety, *Crocodylus porosus,* and one more confined to fresh water, *Crocodylus johnstoni*. Between 1945 and 1971, by one estimate, 110,000 *C. porosus* were killed for their skins, this from a population that in 1945 was said to be in the vicinity of 100,000 (Edine-Brown and Webb 1988). Because of its inferior skin (it has large belly scales), the *johnstoni* crocodile was largely ignored until about 1958. But by 1964, the year the government began protecting *johnstoni,* an estimated 60,000 had been slaughtered in the Territory. The "protection" was more or less in theory, for they continued to be hunted and then sold through markets in Queensland as late as 1972. Once again the territorial government was making it clear that, as with overgrazing and feral horses, it was far more interested in generating income than in protecting wildlife. The more highly valued *porosus* received no protection in the Northern Territory until 1971. Not until 1984 was the export of skins forbidden, even though *porosus* was internationally listed as an endangered species beginning in 1979. In 1985 the status of *porosus* was changed from endangered to threatened, where it joined *johnstoni,* thereby opening the door for "ranching" (or farming) based on a concept of sustainable yield (Tropical Resource Management Party 1988).

The ecologist returned with news, and it wasn't what I wanted to hear. Rangers and ecologists to the west and south in Western Australia said that the most they had seen in recent weeks was thirty to forty zebra finches, and those were dispersed. A few here, a few there, that was it. They were also all redbeaks, which meant that breeding was slow, maybe nonexistent.

I briefly entertained a preposterous thought: zebra finches are endangered in the Outback; to bring back viable populations, we'll have to import them from Nancy's lab, where we regularly joke that they breed like raindrops!

At Keep River National Park I was greeted by Neils, the Danish engineer who, Heidi had said, was formally in charge of her Gouldian finch team. Blond and fair, every bit the Scandinavian, Neils wore thick glasses and spoke with a heavy accent. He had been in Australia for four years, most of it spent with an engineering firm in Sydney. This, he made clear, was a minor accomplishment, of no particular note alongside the more than one hundred species of native Australian birds he'd added to his Life List since arriving. Even this paled in comparison with his consuming passion of the moment: making a salable film on the endangered and truly gorgeous Gouldian finch, a film that he hoped to sell all over Europe—and possibly in America too. He wasn't so sure about that, not having been to the States.

After a brief display of earnestness about what he'd heard from Heidi relative to our project and an overview of his and Terry's daily routine, Neils said he had only the vaguest idea of the purpose behind the data they were collecting for Heidi. What did I think?

"Give me some time to see what kind of data you're collecting," I said.

He showed me around the porch, then to where he and Terry and the other two volunteers, Dean and Henry, camped. He pointed to a couple of cots and said I could choose between them.

As I was settling in, Terry, Dean, and Henry returned from the field. They were exhausted, hungry, and dirty. It had been another long day on nearby Newry Station climbing ladders and carrying them up and down hills checking on Gouldian nests.

Terry and Neils invited me to spend a day with them looking for new Gouldian finch eggs in tree hollows, an exercise that by the time I arrived had resulted in the two of them, and others on Heidi's team, climbing some 450 trees. All of Heidi's intensive research in this part of the Territory was conducted on Newry, a 2,350-square kilometer cattle station whose western boundary was the border between the Northern Territory and Western Australia. Keep River National Park, a triangular-shaped wedge in the northwest corner of Newry, had once been part of the station. Several portions of Newry were deficient in grasses favored by cattle, but the station still ran about ten thousand head in a given year. Heidi had been drawn to Newry because of the many snappy gums or *Eucalyptus brevifolia* on the station, trees notable for the termite-fashioned hollows in branches and elbows that are favored as nesting sites by both long-tailed and Gouldian finches. Presumably, the deeper and narrower the hollows the better, for they offer protection against predatory birds, snakes, and the elements.

The principal site on Newry where Heidi did her survey work on Gouldian nesting behavior, and rather incidentally on long-tailed finches, was some twenty kilometers east of park headquarters. Beginning a couple of hundred meters to the south of the Victoria Highway, the site is a rocky and intermittently hilly parkland of scattered but abundant snappy gums. The sharp rocks and large boulders—often black from burning—are surrounded and covered with lots of prickly spinifex and with sorghum grass. Before the dry season takes hold, the sorghum grass may reach more than two meters in height. Its seeds are favored not just by the Gouldian and long-tailed finches, but also by doves and other birds.

By midafternoon we had climbed ladders on more than thirty snappy gums, looked in more than four times that many hollows. Some trees had numbered metal tags on them, others didn't. We searched the snappy gums on parts of one hillside, then ignored the rest of the gum trees on the same hillside. Memory, it seemed, was the prevailing guide for what to do next and where to go.

Terry and Neils had no data sheets for their fieldwork. Rather, each kept his own notebook, and in the field protected it as if it were an intimate diary. And, just as if going to a well-concealed tryst, each headed off in his own direction with nary a word to the other about what he was going to do. Eager to be polite and unobtrusive, I said nothing. But I couldn't help think, Was this a new-age way to do science in the field, or was there something awfully rickety about this project, in the most fundamental sense?

By day's end, we had found a grand total of nine eggs in one longtail nest and four in another. We had seen two male Gouldians, one proud and hungry-looking brown falcon, and a small bat in one nest thought to belong to a nesting Gouldian pair. Late in the day Neils came upon a Gouldian nest with four young that were ready to fledge. He tracked me down to see if I wanted to watch him band them. We returned to the nest immediately, only to discover that two of the brood were missing. Neils was distraught, even more so when one of the two remaining juveniles escaped from his hand before he could get a band on it. "Heidi will be furious if she finds out," he said.

My focus on longtails, I searched among notebooks that Terry and Neils shared with me. I was looking for some data I could pass on to Nancy (table 1).

I got a letter from Nancy. She wrote:

The students are anxious to go. Even Jean seems more interested now that we're talking longtails. Nat has lots of anxiety (what should I do with my apartment? what's my thesis going to be on?), some of which is rubbing me the wrong way. The fingerprinting is going slowly, lots of equipment problems, but some of it is Nat's fault.

Cole is a live wire and he hates being restrained. Without Amy I am going to have a hard time getting anything done. Amy is quite allergic to the lab, and I'm not sure whether she'll be able to handle birds regularly.

I talked to Heidi and she sounds very pleased about our working on longtails. I convinced her that we can help her out by doing behavior observations on Gouldian–longtail interactions and possibly by adding nest sites. We would color band under her permit. We

Table 1. Notes on long-tailed finches from the 1991 notebooks of Neils and Terry.

Nest no.	Nest under construction or abandoned	No. eggs in nest	No young in nest
494	X		5
5027			2
5029		2	
1077		1	
5040	X		
1050	X		
229	X		
233		2+	
919	X		
5036	X		
5078	(1 dead bird in nest)		
1034	X		
342		5	
5022	X		
1060	X		
5018	X		
109	X		
5021	X		
Total nests	12	4	2

Approximate number of trees searched: 450
Approximate number of hollows searched: 1,200+

have a Top End permit for research on longtails; getting it to cover more birds should be no problem. I'll do that. You can trap and band anytime.

If you can't find zeebs anywhere and need something to do on longtails, the priorities are (1) select a study site (Keep River sounds best from here, but you be the judge); (2) make a map of nest sites and number the nests; (3) start observations at nests to ascertain parentage. If longtails are easy to catch, I wouldn't bother drawing blood now. If they are difficult and you're spending lots of effort catching them, it would be worth taking blood. Longtails use the same band size as zeebs, so you can use those on hand.

KW is supposed to show me how to lap birds to sex longtails. I don't believe Margaret's spring-breeding hypothesis, but the data base is so lousy it may be that zeebs *prefer* spring. They do breed any time of year in central Australia, according to our records and those of others.

I have thought of a couple of things we can do with longtails/zeebs/Gouldians, but have to work up techniques, and five o'clock in the morning is the wrong time!

Your mom says sorry she doesn't write, but she's thinking about you. I am really, really busy. Cole doesn't let me get much done when we're together and he's awake. I work from 8 A.M. to midnight most nights, and Cole still gets up. He can crawl up steps and open cupboard doors. Got himself out of his stroller strap twice this weekend, fell on his head once.

Am starting to panic about getting everything done, but am looking forward to being in Oz with you. I am sure I will be exhausted and teary when I see you, so don't take it personally. Gotta go. Love, Nancy

Heidi had given Terry and Neils instructions to burn two fifteen-hectare patches on Newry, habitats where the dominant species was snappy gum. One plot was to be burned as soon it was dry enough to do so, the other later in the year, toward the end of the dry season. Heidi wanted to know whether a late-season fire, hotter and more damaging to plant life than an early-season fire because there would be more fuel and the vegetation would be drier, would produce a disproportionate number of multistemmed snappy gums. Late-season fires, Heidi hypothesized, had increased in frequency after the arrival of Europeans in northern Australia. Multistemmed snappy gums, those with more than one trunk, have smaller branches and fewer hollows, and therefore ought to have fewer favorable nesting sites for Gouldians than their single-trunked ancestors.

Aborigines had burned the woodlands of the Top End for tens of thousands of years; according to what could be pieced together from the historical record and from oral histories, the effect, if not the explicit design, of the burning was to create habitat diversity, a mosaic of floral and faunal environments that could be exploited at different times of the year. Because much of the burning was done in the early part of the dry season, trees were not greatly stressed or killed by fire.

Once Europeans, primarily Englishmen, began to use the Top End extensively for grazing cattle, the situation changed. Not only were the British unfamiliar with Aboriginal methods of burning and their rationale, but they were afraid of fire. When Europeans burned to produce fresh green feed for their livestock, they did so indiscriminately with regard to timing. That meant that since many of their fires were started late in the Dry, when highly flammable biomass was abundant, the fires were very hot; they destroyed fruit, killed trees, and changed form and growth characteristics. If cattlemen did not burn at all during the prolonged dry season, the effect might well have been the same, for then fires were started by lightning. Undoubtedly lightning was responsible for numerous fires before the arrival of Europeans, but their effect was probably less widespread because of the diversity and patchiness of the fire-created habitats that had resulted from continuous Aboriginal burning.

Gouldian finches are hollow-nesters and are known to have a strong predilection for two species of eucalyptus, *E. tintinatum* (salmon gum) and *E. brevifolia* (snappy gum). Hollows in the trees are formed by termites. If the branches are not sufficiently large, then either the termites will not find them attractive enough to make hollows or the resultant hollows will be too small for finch nesting. In either case, the outcome is fewer hollows in which Gouldians or other birds that use them—in particular, the long-tailed finch—can nest. It may be that hot or late-season fires, those that have been much more common since white settlement of the Top End, stress *E. brevifolia* sufficiently that many in the next generation are multistemmed.

Terry and one of the rangers were disturbed by Heidi's directive to burn the snappy-gum patches. No thought had been given to how to control the geographic extent or intensity of burning. The plots to be burned were not going to be carefully measured. No inventory had been made of tree sizes and ages and nesting sites. No one had any idea whether nesting sites in the known habitats were abundant or in short supply, either for Gouldians or for longtails. Nothing was known about whether Gouldians and longtails were competing for the best nesting sites, since nesting, competitive interactions, and all other kinds of behavior of both Gouldians and longtails had thus far been ignored. Were conclusions to be drawn after one year of burning or after five or ten? It was far from obvious, assuming that there was a difference between single-stemmed and multistemmed snappy gums, how such a finding was to be reconciled with issues of intensive trapping of Gouldians in the 1980s or with other kinds of habitat destruction. Air-sac mites, native to South Africa, were now thought to be widespread in Gouldian finches, and Heidi was treating all captured birds with a medicine that she gave them orally. Whether air-sac mites were decimating Gouldian populations was unknown.

But Terry and Neils did what they were told, and I went along to watch. I witnessed an effort that, I was certain by the time we'd left, was destined to prove only that, early in the dry season in patchy sorghum grass, ten or twelve matches were required to burn a square measuring two and a half by five meters, whereas in other areas thick with shrubs and accumulated dry matter, and with the help of a wind, the only sensible thing to do after throwing a match was to run like crazy.

I wrote a long letter to Nancy, with several suggestions on what to bring for work on the longtails. I mailed the letter, then drove over to the Western Australian Conservation and Land Management Office in Kununurra to meet with Bob Taylor, one of the principal rangers at Bungle Bungle National Park. On the phone he'd said that he had a bird list of 164 species found in the park and, to my utter delight, that he was "absolutely certain" that zebra finches were breeding. He knew where; he could take me to them. He had just returned from the park, where he'd seen the active nests with his own eyes. If I came by the office, he'd give me more details and we could make arrangements for a trip to Bungle Bungle. I would be his guest.

Finally! I thought.

Bob Taylor had other business and couldn't go to Bungle Bungle anytime soon, but the district manager, Mark Pittivano, would be going to the park. First he had to spend a day in Halls Creek, attending a meeting of pastoralists and farmers. The purpose was to discuss some research findings of the Western Australian Minister of Agriculture on the causes of the dramatic increase in siltation of Lake Argyle. Did I want to go, knowing there would be a daylong detour to Halls Creek?

Truculent as ever, pastoralists and farmers wanted no part of what they heard at the meeting, or from the Minister of Agriculture. Not surprisingly, the results of the government's study pointed to overgrazing as a principal cause of gullying and siltation, one major effect of which was to shorten the life of the huge and very expensive dam on Lake Argyle. Any admission of guilt on the part of the pastoralists, or acceptance of the government's findings, would result in loss of control over their properties and destinies. And that they would not stand for, they shouted.

Clive Stone, a brassy and arrogant station manager whom I knew from our previous visit in 1986, sat hunched in a rear corner throughout most of the meeting. Then he saw an opening. Standing as tall as his short frame would allow, he said there was absolutely no doubt in his mind that heavy siltation of Lake Argyle was caused by the torrential downpours that came during the six-month Wet. Everyone knew that, Clive said.

Everyone also knows—no contradiction intended—that station managers like Stone have no choice but to allow cattle to pile up in huge numbers in bad years, at those few choice riparian locations where feed and water are abundant. The managers have their heads firmly caught between the jaws of an ever-tightening vise. Their job is to produce profits for the absentee owners who employ them. If that means letting cattle eat out the land, then that is the way things must be.

Clive Stone wasn't the only one with his head in the absentee landlord's vise; 70 percent of the cattle stations in this vast quarter of Western Australia known as the Kimberleys were run by managers whose masters only wanted to know the size of the yearly profit. Absentee landlords almost everywhere, their focus on short-term profits, rarely thought about long-term range conservation and management issues.

I spent some time traversing Bungle Bungle National Park with an Aboriginal ranger whom I knew only as Paul. He had worked in Bungle Bungle for two years. His charge was to take me around the park and show me zebra finches and where they were breeding. He took me everywhere. Unfortunately, Paul knew little about finches or their habitats, and he didn't recognize the names of trees I mentioned where we might find their nests.

I was relieved when Bob Taylor arrived; my search would soon be over. And over it soon was. We drove to the ranger station, and there Bob revealed that the breeding zebra finches he had referred to were in the eaves of the long rectangular roof. He pointed out the nests. I climbed a ladder, stuck my hand inside, and pulled out hatchling finches. Momentarily I exulted. I examined the hatchlings, then looked around and caught a glimpse of the parents watching the goings-on from a safe distance. They were longtails.

Was this, I mused, an omen? An omen that said that however intent Nancy was on studying zebra finches, it was preordained that we were fated to spend the next several months working with longtails?

I received another long letter from Nancy. The tents for us and the students, on order for some time, still hadn't arrived. Also, for inexplicable reasons (especially given the tens of thousands of dollars we were bringing into Australia), the Australian consulate still hadn't given Nancy and Cole the six-month visas requested.

Nancy hadn't heard from Jean in three weeks, and when she did, Jean reported that she'd been in an automobile accident. She had received only minor injuries, which she said wouldn't affect her performance in the field.

I had a couple of weeks before Nancy arrived, and I could either begin collecting longtail data and preparing the campsite on Newry Station where we'd all be staying, or I could get back on the road and make one more major effort to find a decent-sized colony of breeding zebra finches. Ignoring the omen that flashed before me in Bungle Bungle, I checked over the van, made a few phone calls, and mapped out a route that would take me east to Borroloola on the Gulf of Carpentaria, south to Winton and sites that had shown a modicum of promise, and then east on a long winding loop through southern Queensland, New South Wales, and South Australia before heading north into the Northern Territory and familiar countryside around Alice Springs.

My first stop was Edith Falls. After a relaxed introduction to Heidi's eastern team of Gouldian chasers, I learned that the ecology of Gouldian finches here was somewhat different from that to the west on Newry Station. On Newry the Gouldians are found together with longtails. But the longtails are strikingly unlike their presumed cousins to the west, in that they sport bright red beaks. I wondered: Would they breed with their yellow-beaked relatives? Was there a smooth transition in beak color between the yellow-beaked longtails at Newry and those found at Edith Falls? Or was the beak color change among populations abrupt, suggesting that they had long been isolated, and for reasons probably not obvious had been subjected to ecological pressures that resulted in morphological differences signaling an unwillingness to interbreed? These were certainly questions that would intrigue Nancy, and for which she'd want to gather data.

I spent a long day with Heidi. I was fascinated with her avidity for setting fires in six-foot-high sorghum grass. She burned grass to try to prevent conflagrations that could easily consume hundreds or thousands of hectares of land before containment maneuvers could even begin. The Gouldians, Heidi claimed, were largely indifferent to hot flames. The fires could be scorching trees to a height of six meters or so, and that wouldn't be enough to get the colorful hole-nesters to abandon the young they were incubating.

Between repetitious bouts of expertly lit and tossed matches, Heidi revealed that she had wanted to get the status of the Gouldian finch changed from threatened to endangered. But her bosses at the Conservation Commission had urged that she proceed with caution because a gold-mining company was about to set up operations in the midst of prime Gouldian habitat. The Conservation Commission, Heidi didn't say—didn't have to say—was intent on keeping to its real mission: resource exploitation.

At Daly Waters, before turning east onto the Carpentaria Highway, where I could expect to drive for six or seven hours and maybe see one car or truck, I discovered that for a mere $A10 I could get crocodile insurance. The Life Benefit Crocodile Policy stipulated that were I to be grabbed by either species within five kilometers of the coast, and if I died from this morally innocent taking within thirty days, Nancy or whoever I named as beneficiary would be better off by $A50,000. I decided it was a waste of money. Surely I wasn't stupid enough to find myself in a river full of man-eating crocodiles.

With no luck on finding breeding zebra finches anywhere around Borroloola, I drove south four hundred kilometers on the desolate Tablelands Highway, then on to Winton for a second look at places that had shown a smidgen of promise. But the new search turned out to be worse than the first. Two bushes and one breeding pair at one site; two breeding pairs among twoscore trees and bushes at another site.

On leaving Winton, I headed southeast into country that was naked, habitat that zebra finches don't like, with lots of kangaroo road kills. In Longreach I stopped at a Commonwealth bank and withdrew several thousand dollars, and happened on the best vegetarian sandwich in memory. Then I paid a visit to the region's Arid Lands Office to talk with anyone who might know anything about the whereabouts of zebra finches. It proved to be another cipher—except for information of a different kind. The kangaroo population in the region was said to be down some 40 percent, because of an unidentified virus in a sand fly. But no one was worried; the 'roo population was always up and down, a fact confirmed again and again in the historical record. Besides, the farmers thought there were far too many of them, and they couldn't get rid of them. Not easily, anyway. They'd estimate that they had four or five thousand on their property, shoot that many, and then discover that they *still* had about that many hungry beasts to contend with. Little wonder that long ago most farmers adopted the attitude, If it moves, shoot it.

I turned south, in the direction of New South Wales. In a small town called Tambo, a petroleum driller said that in the 1940s west of Newcastle, he got a cent and a half for every pair of zebra finches he caught. His revenue jumped an astonishing 400 percent when he got his hands on a pair of rosellas. Times had changed, were really bad now, he opined. He didn't have a clue where I could zebra finches, in New South Wales or Queensland or anywhere else for that matter.

Well into South Australia and on a southerly beeline for Port Pirie and the Spencer Gulf, I got a sudden urge at an insignificant place named Yunta to get onto a dirt track heading north. I promptly got lost, pulled off to the side of the road for a long snooze to find my mental compass, then continued on, hoping I wouldn't run out of gas or have engine trouble. Coming up over a small ridge, I saw a diffuse cluster of prickly acacia. In one that topped out at about three meters and dared me to enter with bare hands and arms, I counted seven zebra-finch nests. Five of them were active. I was positively elated. Lost or not, I was determined to find another two or three or four dozen. But whether simply luckless, or blind, or both, I could find no more than three additional nests. Frustrated once again, I drove on slowly, then backtracked. Nothing. I concluded that there were no more to be found locally. I'd simply stumbled on a small band of breakaway colonists who had decided to follow that evolutionary imperative: reproduce or go extinct.

I took this find as omen number two, guessed at which dark rolling track would take me west, and somehow found myself in Hawker just in time to be an uninvited but welcome guest at a wedding reception.

The next morning, tracking north to Leigh Creek, I kept seeing one public banquet after another: wedge-tailed eagles and foxes and cats feeding on the carrion of cattle, rabbits, 'roos, and emu. The food chain, building mass and numbers somewhere near the top. From a chat with wildlife rangers I learned that to the east of me in the Gammon Ranges and Flinders Ranges some twenty thousand feral goats ranged at will. Despite protestations of local Aborigines, who wanted to harvest the wild goats slowly at the going market rate of five dollars a head, the goats had to be cleaned out sooner rather than later. They were causing that much havoc to plant life. They were "positively killing the land," I was informed. The Aborigines had a rightful claim on the goats, but not on the land. By any measure in the mind of a half-sane conservationist, the land is far more valuable, and more difficult to recover, than any number of gone-wild goats (or wild donkeys that wildlife rangers were shooting surreptitiously). There was no need trying to explain to the public an issue burdened with emotion and precious little sense in terms of what really matters.

I wandered into a South Australian killing field. Out on the Strzelecki Track, a "place to go to find zebra finches" (or so my latest source claimed), I found little. Unable to control my snoop genes, I detoured onto a sheep station. Five young muscular fellows half my age were in the process of shearing eight hundred sheep. Each of them could clip as many as 160 a day and would get $A1.36 for each one denuded. But the real news wasn't about numbers sheared and money made and back and muscle strain, it was about seemingly senseless death. The station would be getting rid of seven thousand of its thirty thousand head of sheep. All those I'd seen shorn were taken out to a ditch and shot and pushed into a mass grave. On the station to the west, conditions dictated that ten thousand of twenty-five thousand be shot. On yet another station not far distant, the killing would reach thirty-five thousand this year alone. The stations were keeping the breeding rams, so they expected to be shooting again next year. The rationale for all this killing was simple: too many sheep, and the government was paying more for a carcass in a ditch than the owners could get for the meat at market.

I still had a chunk of northern South Australia and a piece of the southern reach of the Northern Territory to explore. But by now I had seen enough, and had enough omens, to convince me that short of a chancy miracle, Nancy and I and our team of students would be spending our time with Gouldian and long-tailed finches. Mostly longtails, or, as the lingo would have it, blackhearts.

3
Mysterious Behavior

Nancy is happy to be in Australia and eager to get to work. But her elation at having arrived can't hide obvious signs of stress from the previous months: too many demanding university committees, selling our home and storing personal effects, getting the lab prepared for her departure, quieting the students' panicky last-minute needs, the logistics of all that had to be rethought because of the change in research plans, and not least Cole's increasing demands, which have been compounded by the string of ear infections that began before I left. Cole initially wants nothing to do with me, and who can blame him. I'm a complete stranger, and I insist on hugging him to death.

To help Nancy get over her jet-lag and tiredness, we get a pleasant motel room in Darwin for two nights. We spend our time getting reacquainted, eating Greek and Lebanese food, going through my detailed notebooks, wandering through a museum admiring Aboriginal art, looking for a sextant that had been on the *Beagle* during Darwin's famous voyage. The sextant, we learn from a starry-eyed museum employee, has mysteriously "gone missing."

In Katherine, and then in Timber Creek, we spend hours with Heidi over dinner and drinks, outlining our plans, and giving assurances that anything we do at her field sites won't disrupt her Gouldian project. Nancy says she has several broad objectives: gathering data on longtail reproductive behavior; learning what she can about competition between Gouldians and longtails; taking measurements to determine degree of left-right asymmetry of body parts and plumage; and collecting data on beak color and related phenomena that should prove useful when conducting sexual selection experiments in captive populations. Much more is on Nancy's mind as well, but this is enough to give Heidi a grasp of what we're after.

By any measure, Heidi is generous, and altogether enthusiastic about our working on blackhearts alongside her and her team of volunteers. She confesses to seeing the considerable payoff of our involvement with her Gouldian research. More Gouldians will be caught and banded and thereby more data will accrue. By focusing on longtail nesting and breeding behavior, Heidi may well get new insights into the central question of Gouldian decline. Not least is the advantage of having Nancy around to discuss hypotheses and techniques of data gathering and analysis.

I'm somewhat surprised at how excited Nancy has become about the prospect of studying longtails. "How come?" I ask.

She says: "The Top End climate is much more predictable than that of central Australia. Perhaps, unlike zebra finches, this species is sedentary enough to be a good field species for me. Maybe, in the long run, I could even run field experiments on longtails. If we pick a decent site and work it intensively, and also come back next year, we can get a good idea of how much turnover there is within and between seasons. From what Heidi says, we can at least count on the longtails being at the same places next year. In addition, we can do short censuses at other places in the Northern Territory, and maybe even Western Australia, to examine variation in beak color and plumage markings of longtails over their range."

Upon arrival at Newry Station, our first task is to clean up the site that Alan Andrews, the manager, has agreed to let us use for the token sum of $A40 a week. Alan is responsible for the day-to-day operations of Newry, for ensuring a profit for the absentee owner who lives in Brisbane and rarely visits this and other stations that he owns in the Territory. Although Alan knows little about biology and the wildlife on Newry, he likes the idea of having researchers around. Perhaps we'll teach him something of what we learn about the local finches.

We will be far enough from the main set of buildings (which includes the family home and bunkhouses for the resident ringers, a garage, and a generating plant) to cause Alan no obvious problem. Although he hasn't said so, I suspect he sees us as a geographic barrier between his family and the unknown traffic on the Victoria Highway, a hundred meters or so from where we'll camp. The unspoken premise is that predatory highwaymen would hit us first, and either proceed no farther or give Alan and his family enough time to get out of harm's way.

The site where we'll put up our tents and eat our meals and process data is known as Center Camp. When Alan first showed the camp to me, he described it as "flash." This I understood to mean that by outback standards of the time—the camp was in use from the 1940s until 1981—it was about as upscale as anyone could find in remote Australia.

The camp is principally composed of two metal buildings with corrugated roofs and cement floors. The walls are made of sheet metal and lots of window screen, which is old and peppered with holes and large gashes that make it easy for insects to come and go at will. The larger of the two buildings is now used for storing hay. The smaller one, which we'll use, is about the size of a small efficiency apartment. It had once been a kitchen, a cook's quarter, and an eating area for white ringers. Aboriginal ringers—then certified outcasts even to the federal government—got their food through a "dog window," a small square hole near the sink. The hole has no covering and is large enough for birds and bats and other small mammals to enter.

Inside the tiny building is a small squarish room that had been a cook's bedroom. At the front end is a tiny cement-floored shower, now masked by a plastic green curtain I recently bought. In the cooking and eating area, which takes up most of the building's space, there's a small sink, several rudimentary metal cabinets for storing food, a wood-burning stove on top of which I've put a gas camping stove, and a long picnic-like bench where we can eat and process data and sort birds that we bring in from the field. For a toilet, we'll use one of the two primitive outhouses that sit between the cook's house and the ravine through which flows the often-dry Keep River. One of the outhouses has "Colts" written on it, the other "Fillies." The names are an artifact of a time when those living and working on the station held occasional soccer matches. Only the one labeled "Colts" is in good enough condition to use. At the moment I don't see any need to repair the other one.

Near one corner of the building that will be the focus of our life when we're not looking for or trapping finches is a large water tank on giant stilts. When full it holds about a thousand liters. The tank is open on top, and a pipe extends about halfway across. Because the pipe is a perch for the hundreds of little corellas that live in this section of the station, the water is constantly being spiced with cockatoo excrement. So far I've not had any reactions from drinking the untreated water, and we'll proceed on the assumption that processing a little cockatoo bacteria in our own plumbing is nothing to worry about.

The only other demand for this water comes from the thousands of cattle that still move through the complex of corrals (*yards,* in Australian lingo) that are a mere fifty meters or so from Center Camp. We're guaranteed of not only seeing and hearing thousands of nervous penned livestock at all hours of the day, but also encountering numerous wallabies and station horses that roam through the area day and night.

Soon the first major herd of the station's ten thousand or so head of cattle will be pushed through the main station gate on horseback and brought into the complex of yards, to be sorted, branded, dehorned, and checked for tuberculosis and brucellosis. Marketable steers, and some cows, will be trucked to the port of Wyndham to the west and from there shipped live to Indonesia.

Nancy and I spend the better part of two days cleaning the floor, the walls, the cabinets, the stove, the shower, everything: of dirt and grime and beer and bean cans, spiders and long-dead lizards, piles of old papers, and fetid and greasy rags from another era. Nancy is principally concerned with the peeling paint, which she's certain is lead based. It would be all right to ignore it were not Cole with us. But Nancy is vividly imagining how much he can peel off and eat, and that irreversible brain damage will result. I'm afraid to count how many hours she has already spent on her knees scraping and brushing and sweeping to soothe her anxiety. She announces that we should plug his mouth with a pacifier whenever he's on the floor of the building.

After we fill the cabinets with staples and make a few minor repairs, still feeling full of energy we open a bottle of chardonnay. We celebrate the formal beginning of fieldwork. On a whim, thinking of the finches that are about to become Nancy's new study organism, a potentially great source of data to compare with what she's discovered about zebra finches, I christen our field abode the Blackheart Hotel.

We begin a systematic survey of the field sites. This is pure fun, not least because I again take it on myself to get into the naming game. I christen this place and that place where we'll trap with mistnets or use the small walk-in traps we made in Illinois. It's a playful time, and I tease Nancy by naming trap sites Arlette, Chloe, Belinda, and Tanya. I claim that they're old and dear friends about whom she really doesn't want to know anything more than what's revealed on my face. To prove that I'm not thoroughly sexist, I name a couple of promising field sites after songs by John Williamson, a well-known Australian folk singer: Mallee Boy, Cootamundra, True Blue.

In Kununurra, we can't find nearly enough of the kinds of pureed vegetables and fruit we want for Cole, so I call Amy and tell her to bring as many different kinds as she can fit in her bags. "And don't forget the international adapters for my laptop," I add. "I'm going insane keeping my journal up to date in longhand."

Nancy meets Terry and Neils, and we accompany them into the field where she quickly assesses their data collection methods. The scientifically indefensible methods make her irritable. "Pseudodata," she scoffs in private. But it's a problem that can be ignored. After all, we'll be collecting and recording our own data, and we don't expect to be critically dependent on anything done by Heidi and her volunteers.

A couple of days before our students arrive, we make some maps and rope off several study plots and clear large boulders—those we can move—from the poorly graded tracks we'll be rolling over daily in the van. Through all this Cole jabbers endlessly, or naps. And when his face takes on a mottled charcoal appearance and the bush flies pile up like a freeway crash scene at the corners of his eyes, he behaves exactly like Aboriginal children—with complete indifference. For pleasure, pure pleasure his laughter proclaims, he likes nothing more than sloshing around in a tiny pink plastic wash bowl that we place in the middle of the kitchen floor at the "hotel." Happy to see everything moving along so smoothly, every time I see Nancy getting fastidious about keeping the cement slab clean I tell her it's a ballroom, grand enough for dancing. Sarcastically, she reminds me that she's already found several red-backed spiders that she'd heard are not exactly harmless, especially for a child not yet a year old.

When we repair to our double Australian swag at night, Cole firmly and resolutely wedged between us, we talk with high anticipation about the arrival of the students and all that we'll accomplish. We're all but certain that an unparalleled experience has just begun.

I'm late to their bus, and they've left the hotel, where they were deposited, for an unnamed hostel. It takes me nearly an hour to find them. They're road weary, tired of buses. "It was the Bus Trip from Hell," Jean says several times. The Bus Trip from Hell—more than halfway across the top half of the Australian continent—incurred tangible losses. Tim hadn't paid much attention to the luggage, and one of his bags and an expensive six-person tent we'd bought for the students disappeared somewhere en route. I say that these things happen all the time; we'll return in the morning and see if we can put a tracer on the bag and the tent.

Heidi's team of volunteers isn't doing any trapping or Gouldian nest searches at Dingo Creek, so it looks as though we'll have that site to ourselves. Trapping without Heidi's team in these first days will give us a chance to refine our methods of trapping here. We won't be distracted in these initial efforts at banding, recording data, drawing blood, weighing birds, measuring wing length and body parts that Nancy judges worth the effort—data points for hypotheses that thread through her mind like veins of fat in feedlot beef.

We decide that we'll trap at Dingo Creek on days when not gathering behavioral or ecological data, or we'll mistnet-trap at either Tanya or Chloe, the two artificial water sites west of Gouldian Mountain that Neils and Terry faithfully tend for Heidi. Water at these sites is trucked in monthly or as needed. It is fed into large red watering devices made for poultry, which finches and other birds have adapted to as if they were natural water sources. If we trap at Tanya and Chloe on those days when Neils and Terry are not running up and down ladders looking for Gouldian nests, their assistance will allow us to process the birds more quickly, decrease the likelihood of losses, and perhaps get additional data that are time-consuming to collect.

The blackhearts are plentiful at Dingo Creek, and the long pool of muddy water that sits in the depression in a bend of the dry creek is easy to trap. We encircle the open edges of the pool with two mistnets, one eleven meters and the other six meters in length. Each, when stretched vertically on the staked poles, and allowing for five ample pockets into which the birds fall upon hitting the nets, reaches to nearly two meters. Between the nets and the water we leave small "beaches" on which the finches can alight. From these small dirt patches they head for the water—if they haven't hit the nets and dropped into a pocket on their sharp descent. Still more birds end up in the pockets of the nets when the finches attempt to return to the nearby trees.

We have trouble getting enough "bends" in one of the nets, because part of it extends over two centimeters or so of water. With several birds in the lowest pocket, they weight it down enough for the finches to get wet. If we're not attentive to getting them out quickly, they can drown. We go over this possibility with the students, remind them that birds nearest the water are our first priority. Those in direct sunlight or badly tangled are second on the list.

Alan has some cows pasturing on the bluff above and behind the pool we're trapping, and now and again one or two of them wander into the broad gravelly creek bed. We've also seen wallabies approach within twenty meters of the water, only to flee when we rush the nets to flush the birds. The cows and the wallabies leave us alone while we trap. But one night we left the white poles standing and the nets rolled up, and when we returned in the morning they'd all been knocked down. The dirty nets took us a couple of hours to clean. We won't make that mistake again.

Above the vertical six-meter bluff that forms one long side of the pool are three middle-aged leafy eucalypts that the finches come to before deciding that it's safe to drop down to water. Depending on wind and the presence of predators (hawks primarily) these trees, and a couple of others some fifteen meters farther up the creek, fill up with as many as two hundred finches at a time. Initially, they come in small numbers—three, five, seven—and I wonder if they're either experienced "scouts" or young birds full of reckless daring. They sing, they flit among the branches, they search the surroundings for predators with their keen eyes. And then, for no reason obvious to a human, one or two adventuresome birds fly down to the water and take a couple of sips before returning to a tree. If there seems to be no obvious threat or hungry-eyed predator nearby—losing one or two finches into a net pocket doesn't appear to count as a problem in their assessments—then shortly finches began coming in waves. Five or ten or twenty at a time, sometimes wave on wave. What an enchanting sight! What a rush! And to a cold-hearted scientist, what a flood of data.

Tim and Jean are surprisingly distant, as if Nancy and I are strangers. They're also listless and show only a modicum of interest in the field sites, what we're doing, the station—everything, it seems. Jean, as I'd anticipated in my darker moments before leaving Illinois, studiously ignores Cole. She's much more interested in the battalions of ants that endlessly troop back and forth on the shattered cement ground beneath the veranda of the hotel.

The two students have been sleeping in Jean's small tent since they arrived. They go to bed early and zip the flap tight.

"Are they romantically involved?" I ask Nancy as we lie in our swag, Cole fast asleep between us. "And do you think that's the source of their distance, their coolness toward us?"

"No idea," she says.

"Let's hope this doesn't prove to be a problem when Amy arrives. Boyfriend-girlfriend divorces may be the norm, but this is the wrong time and place."

We send Jean and Tim to Cootamundra, our westernmost field site. They return, claiming to have identified four pairs of longtails. They have also found a

Gouldian nest that was missed by Terry and Neils in their tree searches. Nancy looks over Tim's field notes. She's not thrilled with what she reads, with the level of Tim's apparent observational skills, his inattention to detail. Still, she thinks she can teach him how to improve. Jean's notes are much better.

After finishing fieldwork for the day, I take Tim and Jean into Kununurra to buy food and other supplies, fill up on gas, find a mirror for the hotel. We've been feeding them since their arrival, and also doing all the cooking. We wanted to ease their entry into this alien, less than luxurious environment, bring them out of their inexplicable funk. But now it's time for them to do their share.

Tim and Jean spend a great deal of time in the supermarket, planning their meals on the spot, scrutinizing the ingredients, and trying to be as economical as possible.

Jean says to me, "I haven't seen any land abuse anywhere around Newry. Or on the long trip from Cairns, for that matter. I thought you told me there was lots of land abuse in outback Australia."

I ask her if she recognizes invader species, plants that are almost certain indicators that there's been overgrazing. I name several, giving her both common and scientific names. I tell her what the plants look like and say that they're common along the Victoria Highway. "You couldn't have missed them on the trip west," I say.

She snickers and walks away.

I wake to the deafening crescendo of little corellas. I roll over and kiss Nancy, check on Cole's whereabouts at the foot of the swag. I cover him up for the fourth or fifth time since he fell asleep. The sweet smell of grass and the aroma of horses that doze just outside the tent fill my nostrils as I glance at my watch and see that it's exactly twenty minutes to seven. The grating screeches of the little corellas haven't varied by a minute in the last five days. Sample-conscious scientist that I sometimes pretend to be, I conclude that there'll be no further need to bring my left wrist to my blind eyes when I hear the racket.

Few finches come to Dingo Creek this morning. Eight or ten doves and nothing else, not even honeyeaters. At first I think the two hawks that sit high in the dead eucalypt across from the water are responsible. Then when they leave, the wind comes up, and this, I surmise, is not something the finches want to fight even with the predators gone. By late morning the wind has become an idling breeze, the temperature is in the nineties, there is little shade around the water hole, and still no birds. *Where* are they drinking? And why this apparent change of behavior?

On previous mornings brown goshawks have come and gone, the wind has made huge pockets in the mistnets, and the temperature has been even higher—yet the longtails and Gouldians and the charming little double-bar finches have come to watch, to sing, to drink, in both small and large flocks. Oh, my ignorance! How little I know about the gathering and drinking and mating behavior of all these marvelous flocking finches, the nitty-gritty ecology of this bird-rich environment.

The listlessness, the distance, the disinterest in our research of Jean and Tim show no sign of waning. Nancy and I are confused, puzzled, at a loss for a reasonable explanation of their unexpected behavior. We're reluctant to approach them on these matters. We're waiting for them to reveal their anxieties and tell us what's wrong.

Nancy is also taken aback by Tim's cavalier attitude about working in the field. He came without a hat and he won't carry water. I've told him and Jean that the sun is our principal enemy, and dehydration isn't far behind. Tim also came without a field notebook, and several times he's asked Nancy or me for something to write with. He also came without a watch. Nancy had asked him to get one before coming, saying that we couldn't send him off on his own looking for and monitoring nests if he didn't know the time. "How can you arrange to meet us at a certain time without a watch?" she asked him. "We're all dependent on one another." Nancy has now told Tim two or three times since his arrival to get a cheap watch, anything that's more accurate than venturing a wild guess by looking at the angle of the Southern Hemisphere sun. He ignores her, shrugs, and laughs with that engaging smile of his.

"Buy him a watch and take it out of his next paycheck," I told Nancy last night.

"No way," she said.

The principle is so basic, and she is so piqued by his attitude, that I'll bet she wouldn't give him one if she owned a jewelry story full of overstocked throwaways.

Roz, Alan's wife, says that Nathan, at fifteen the eldest of their children, is at boarding school in Sydney. The other two kids get their education at home, sitting in front of a radio, their mom between them. I try but can't imagine a situation where either Nancy or I would even contemplate a boarding school for Cole, much less one that's half a continent away. Our pleasure comes from having him with us. Nancy already talks about wanting Cole close to us even when he's in college.

Jean and I drive out to Bostock's Bore on the south edge of the station, looking for a new trap site. We could use another one to take some of the pressure off Dingo Creek, Chloe, and Tanya. If we could catch enough longtails this far from the other water sites, it would provide us with suggestive data on how habituated the finches are in going to water, the distances they're traveling. And who knows what else we might learn?

Bostock's Bore is another of these quite familiar "sacrifice areas," a parched and uninviting landscape mottled with cattle excrement, bones and boulders, anemic prickly acacia, and lethargic cows. I stare and bring to mind my fragmentary images of the Sudan, Ethiopia, the forbidden margins of the Sahara. Only the human misery is missing.

With binoculars I catch sight of nine blackhearts, Jean one. Given the distance, what we don't see, and the difficulty of trapping here because of the long pool of water that the birds use, the site is not promising. Still, it may be worth a morning's effort if bird traffic slows at the other sites.

Jean sees a small flock of zebra finches, a reminder of home and Eddie, she remarks. Her eyes fill with tears. I turn away, at a loss for words, unsure of how to comfort her. She says she fears that in twenty years she won't remember Eddie; maybe she won't be able to describe his facial features or more than a handful of memorable times they had together. Will she focus only on his final days struggling up and down the hospital hallway, totally bereft of the energy that once brought so much light to his face, the smiles and laughter of friends when he told a bawdy joke?

What has really called Eddie to mind, Jean confesses, is his *Field Guide to the Birds of Australia*, which she has brought along. Whenever he saw a new species, he'd write down the date and location beside its picture. Jean has religiously been doing the same.

On the dusty and bumpy drive into Chloe, Tim and Jean once again announce that they're tired all the time.

"Is something physically wrong?" I say.

"It's the jet lag," Jean says.

Tim nods, sheepishly. The now-familiar acquiescing head bob follows whatever Jean says.

You're kidding, I think. They've been in Australia more than a week. I bite my tongue to keep from saying what's on my mind. I'm as confused as ever about their behavior, a confusion that grows with these kinds of explanations.

At Chloe and Tanya, where we trap with Neils and Terry, and rarely with Heidi (who's been spending most of her time in Katherine or Edith Falls—her other major Gouldian site), finches only come in because of the water that Heidi supplies. Once she stops, finches and all the other birds—doves, honeyeaters, the occasional budgie—will be forced to go elsewhere. In an important sense, then, Heidi has made it appear that thriving Gouldian populations exist on this small patch of Newry Station. But without the artificial water supply, it's questionable how many blackhearts and Gouldians would be here, or nesting nearby. Would there be any?

At both Chloe and Tanya we use two six-meter mistnets set up to form V's. The one at Chloe is relatively "closed" (the V is "flat"), and it slopes slightly downhill and away from a large bulbous snappy gum that the birds come to before dropping down to the water container set inside the V and nearest to the closed end. In somewhat striking contrast to Dingo Creek, diamond doves (*Geopelia cuneata*) and peaceful doves (*Geopelia placida*) are often the first to arrive. Their numbers build up to a dozen or so, then longtails and Gouldians make an appearance. Soon they greatly outnumber the doves. We get several batches of what we've after, and by ten or eleven in the morning, if we're lucky, we've bagged and boxed sixty or seventy blackhearts and four or five Gouldians. After that we see no more than a score or so in the trees, and half this many venture down to the ground for water. By noon, trapping success is no better than that experienced by the proverbial freshwater fisherman at this time of day.

For taking measurements of the birds at Chloe, we've set up a small card table about fifteen meters away from the trap, on a small knoll under trees that shade us from the sun. Here, as at Dingo Creek and Tanya, one of us—usually me—agrees to watch the birds build up around the water container. When the pockets have enough birds to keep us busy or when we judge that they'll start to get too hot, I motion and a couple of us run to the nets with waving hands and a child's hoot to scare those on the ground or drinking. All of this fanfare momentarily frightens those birds near the nets enough so that they become disoriented and hit the black mesh as they attempt to flee skyward. Sometimes this tactic adds eight or ten birds to those already in the nets; at other times the effort is completely futile.

At Tanya, the water is just yards from a broad, boulder-strewn station track, in an area that's open and badly scarred from burning and overgrazing. Here a wider V works better that the narrower version we use at Chloe. Though the catch from one trap and trap day to another is highly variable, the proportion of Gouldians caught at Tanya on a given day seems higher—often much higher—than at Chloe or Dingo Creek. This pattern isn't one just any gambler would bet on, and it isn't easily derived from what knowledge of Gouldian nesting patterns we've gotten on our own or gained through Terry and Neils. Nor is it a mystery that reveals itself through purposeful walks in these habitats. We can, of course, only speculate on what we'll be catching, if anything, at these traps a month or six weeks from now.

Whenever we trap and finches are sparse or we see them leaving on a high flight pattern, I make an effort to judge whether they're heading for water at one of the other traps. The few times I've tried to follow up and make sense of these I-know-where-they're-going leavings, I've come up with little in the way of firm conclusions. I decide the effort isn't worthwhile because it takes travel time and another twenty minutes or so at midmorning to set up the nets at a site not chosen that day, and we're losing valuable catch time. The window of opportunity for getting abundant data from trapping is small.

At all of the traps, we quickly get into a rhythm of erecting the nets so that we're sipping coffee and waiting for the finches no later than seven-thirty in the morning, often earlier. One of my jobs is to get everyone out of the sack at five-thirty, by turning on the hotel lights and giving it the old "Let's go, everybody out, double time!"

Nancy, well organized and intent on staying up to date, spends afternoons preoccupied with data: which data to gather, in what order, how to analyze them quickly to assist day-to-day planning. When she's involved like this, I find it best to give her lots of space. If I do say something to her, she only half hears me, or forgets something I thought important. Her focused and singular attention brings to mind familiar quips between us: about how there can only be one thing at a time for the intense, single-minded professor at her workbench, whereas for me life is multimedia, all diversion, noise, with my senses running in different directions at the same time.

I've bought another aluminum ladder, longer than the one I purchased on arriving. This will allow us to get to even the highest nests, and when Nat arrives we can have two teams working at the same time. I also bought another camping stove. We're down to one burner on the other one, and I haven't been able to find a replacement jet for it in Kununurra. It's one of many reminders of the isolation here, living on the edge of a minimalist frontier.

On the way back to the hotel, I pick up an Aboriginal woman and four of her seven children. They're all between four and eight. One of the boys is fascinated with a hand calculator that's lying on the seat. I remember that we have an extra one, so I give it to him. He smiles as only Aboriginal children can smile. He has such gorgeous eyes and deep brown skin. How completely innocent of the low-slung and degrading life he will almost surely lead! The smile is so fetching I want to hug him. My brief reverie is shattered when his sister, age seven, tells me to put on my seat belt. She says it as if I'm the child, she the concerned parent.

Nancy fiddles repeatedly with the hair at her neck, a sure sign that she's edgy.
"What's wrong now?" I say, asking myself as much as her.
"Why are they so lethargic and disinterested? What's wrong with them? What have we done?"
I shake my head to questions that have started to become mantras between us, a preoccupation greater than the logistics of trapping, eating, keeping the vehicle running, or preparing for the next day.
Jean, we've learned, has had two overriding concerns since her arrival. One is the heat, about which she complains frequently. Her carping begins when the midday sun dares to heat the air to a dry, breezy twenty-seven degrees Centigrade. But why? The weather here is like late spring in central Illinois, much more pleasant that the weather she's just left, nothing at all like the very humid and oppressive days of a typical Illinois summer.
She's no less unhappy about the cost of food, believing that prices shouldn't be any higher than what she imagines paying in a Third World country. Then there's the issue of her salary. Before coming, she was quite happy with what we agreed to pay her, knowing that every penny would be more than she would get from most field ecologists for this kind of work. She says, "I didn't realize that what you were paying me would have taxes taken out."
"Your check comes from the university," I say. "It's no different than how you get your money when you're a teaching assistant or at the survey."
Why should expectations and tolerances be so different here? Is the expression of these differences one way in which we measure psychological dislocation—the shock of the new and the foreign?

Heidi, exuberant and full of good cheer, comes to the hotel to discuss several pressing issues. In four days of trapping at Dingo Creek, our team has caught forty-five Gouldians. Only two have previously been color banded. This might mean that Heidi's team has missed quite a few breeding Gouldians in their searches of hollows. Or it could mean that the finches are nesting beyond the areas that her team arbitrarily roped off. Both possibilities are likely. We have also found several active Gouldian nests that were overlooked. We've come upon them simply by being alert to activity in the trees when we work on or near our roped-off survey plots. All this suggests that Heidi might want to revise the definition of her study sites, or perhaps—heresy this, I fear—rethink her arguments for Gouldian numbers and decline.

Of growing interest to Nancy is a sex-ratio question, a preoccupation as old as her doctoral research on pigeons. Shortly after Nancy arrived, Heidi told her that the sex ratio in the Gouldian population was about fifty-fifty. Our data from trapping several dozen Gouldians suggest otherwise. For every four males trapped, we're only finding one female. Although our sample is still small by Nancy's standards, this striking imbalance raises new questions. There might be a shortage of females, for example, which could account for the allegedly small Gouldian population. If there is a shortage, what's causing this peculiarity? Yet another possibility is that the males are accompanying the juveniles on their morning forays for water and the females are coming at other times, after we've taken down the mistnets.

Heidi is trapping only two days once every three to four weeks on Gouldian Mountain. She says she follows this pattern because a roughly even interval of time between trap dates is required by the population estimation model she's using, and because trapping more frequently is not feasible. She opines that if she traps more often the finches will become "trap shy" and our catches will get smaller and smaller. Heidi hasn't shared her population estimation model with us, so we know nothing about the assumptions involved.

Based on our experience with zebra finches in central Australia in the mid-1980s, our inclination is to trap intensively, even though it's time-consuming. Without regular trapping, it proved impossible to get recapture rates exceeding 5 percent. If we have rates this low, we won't have any idea whether we're dealing with a very large population of blackhearts or one that's itinerant, just passing through. We may not know in any event. In an environment where it's possible to locate only a small number of nests, and most fail because of predation, trapping is the best way to track breeding activity.

Nancy knows that with practice she'll be able to determine with considerable accuracy the age of young of the season at their first capture. She could work solely from her knowledge of zebra-finch ontogeny, but she guesses that blackhearts develop more slowly. She's eager to measure their developmental progression. By retrapping young birds frequently, she can get the needed experience quickly enough to backdate and ascertain the beginning of the breeding season—which we missed—and to quantify its ebb and flow fairly accurately. So questions come to her probing mind. Do the longtails start breeding in the Wet, that fairly continuous rainy season that begins in October or November and continues into March? What is the peak of the breeding season? When is the last clutch start date? During the breeding season do finches frequent numerous water holes over a broad spatial scale, or do they rely on one or two? Answers to these questions would give us a jump start on understanding the reproductive ecology of longtails, and give us practical guidelines for future field seasons.

Frequent trapping would also make it possible to distinguish between residents and itinerants. In Alice Springs, Nancy found that many of the birds she caught first were retrapped again and again, and a declining number of those trapped for the first time later in the season were retrapped even once. Looking back on her zebra-finch data, she discovered that the plumage condition of the residents was very poor when she first caught them. Their feather color was more tan than gray, the result of bleaching by the infamous rays of the southern sun. Their flight feathers were ragged, worn, and thin. These finches had not molted their tail, wing, or body feathers for quite a long time. Shortly after our arrival in central Australia, a two-year drought broke, the grasses grew and set seed, and the zebra finches molted into stunning gray birds in prime condition. Interestingly, the feather condition of many of the early itinerants was much better than that of the residents. How far had they traveled? Why had they left their former sites? What were their ultimate destinations? Now Nancy can only dream that she can answer questions such as these for longtails.

Intensive trapping would not only be one of our major sources of data, and a possible guide to where to go next in the research, but it would also take care of our need to band the blackheart population on and around Gouldian Mountain so that we could follow individuals. It's hard to trap breeding finches at the nest, and we worry that if we did so the parents would abandon nesting attempts.

Our plan, then, is to ask Heidi if she'll let us trap more frequently at her artificial water sites. Every four to five days is what we have in mind. Because of the large number of Gouldians we're getting, sure to increase with more trapping, Heidi might also see it to her advantage to divert some of her team's nest-search time to the Dingo Creek area. Were she to do so, her data base would be greatly enriched. The payoff to us is clear. In their search for more breeding Gouldians, Heidi's team will invariably come upon blackheart nests, and birds that we've color banded.

Once again we've run out of water at the hotel, which isn't surprising given that the cattle now in the yards are taking about four hundred to five hundred gallons a day. A short while ago Jean climbed the ladder on the water tank to assure herself that we're indeed out and that it isn't a matter of poor pressure or a clogged line. When she came down from the tank, she said she was disgusted by all the algae and feathers and cockatoo excrement in the bottom of the tank. I told her not to worry, that no one so far has had diarrhea or other obvious intestinal problems. I said that if she's worried she can boil the water or buy bottled water cheaply in town.

The Rec Room, as it's called here, is a small, squarish shed adjacent to the garage and a rundown tennis court. The nets and the playing surface, full of weeds and large cracks, haven't been repaired in years. There are doors on two sides of the Rec Room that I've yet to see closed. Two of the long walls have windows with no glass and torn screens. Inside the Rec Room are three Formica-topped tables, several wooden benches, a long-dead gas stove, and a small refrigerator that rarely has more than a couple of cans of soda and beer, and a plastic bottle or two of botulinum vaccine type C for testing cattle. A tiny color television with poor reception sits on top of the refrigerator.

On the screened wall that faces onto the tennis court is a fat yellow telephone. For everyone staying at Newry (everyone except the manager's family), this is the only communication with the outside world. Alan and Roz have their own phone and a fax machine.

The yellow phone doesn't ring loudly enough to be heard in the kitchen or the bunkhouses or the main house; if no one's in the Rec Room, it often rings twenty or thirty times. It's from this phone that Nancy and I and the students make all our calls when not in town. Or rather, we try to make calls, for as often as not we can't get through; the line goes dead shortly after putting money in, or the person called is barely audible.

Shortly after we arrived at Newry, I'd been told to inform people that the best time to try to reach us was between eight and nine in the evening. But it turns out that this is just about the worst time. The ringers and all the other help on the station eat between seven and eight, and afterward three or four of them spend an hour or so in the Rec Room reading or having a beer or two before going to bed. A couple of them spend a lot of time on the phone.

Mick, one of the young ringers, is the worst of the lot. Night after night he gets on the phone shortly after he's eaten, and he invariably spends the next hour or so hunched on top of a wooden table in the lotus position, whispering to his girlfriend, who's a cook on a nearby station. Mick just plain doesn't care about anyone else. Even when one of the other ringers tells him that he's expecting an urgent call, or needs to make one, Mick invariably responds by lowering his head and then continuing for another fifteen or twenty minutes before hanging up. When he does get off, he leaves the Rec Room abruptly without saying a word to anyone.

Mick isn't the kind of guy anyone is anxious to set straight on this matter of manners. He's got a prizefighter's upper torso and, shirtless as he often is, he looks mean, eager for a fight. When you ask Mick anything, he grunts or scowls. So I, like others, measure out those times when I feel I have no alternative but to tell him I'm tired of waiting for access to the phone, and would he mind.

At present we're getting in about four to five hours of fieldwork per person per day. Yet at lunch Jean and Tim complain of exhaustion, the need for a nap, a long rest in their hammocks. Nancy and I find plenty to do in the afternoons: filling out and checking and rechecking recorded data, organizing supplies, repairing field gear, preparing for the next day. But getting Jean and Tim to participate in these activities is like pulling molars with greasy fingers. Worse, they won't even talk about the morning's work, and they show no interest in the next day's activities or what lies ahead.

We felt we had no other choice than to sit with Tim and Jean in the hotel after dinner and ask them what was wrong. Nancy, more diplomatic than I tend to be, explained that we were utterly bewildered by their lack of energy, interest, curiosity, involvement in almost everything we're doing. She said, and then I said, that if something was wrong and they could explain it, we'd try to help.

The reaction was one of surprise. "What are you talking about?" Jean said. "It's the jet lag." Her tone suggested that we were making much ado about nothing, and what right did we have to be asking these kinds of questions. Tim, consistent with the way he had behaved since arriving, once again deferred to Jean. He said nothing. She's seven years older than Tim; he treats her as a no-nonsense parent.

We tried—fruitlessly—to get them to open up. There was nothing to talk about, Jean said.

Nancy and I left, feeling frustrated. We sense we have resolved nothing, perhaps made things somewhat worse by letting them know that we're aware of their distance, and indicating our expectations. What can we do?

Amy arrived this evening, running down the road from the station gate, shouting, "I'm here!" A huge smile covered her face as she came up to the campfire where we were finishing dinner. I asked her if she wanted to try some of my outback specials. This night it was yams that I'd put in aluminum foil and buried in hot coals, and another concoction of carrots, potatoes, and large onion slices, well spiced. I'd also made some thoroughly toasted corn on the cob. Amy was game. She dove into the overcooked grub as if she hadn't eaten since leaving Champaign.

Even before Amy began eating, she asked for Cole. She wanted to know everything he'd done since she'd seen him last. She cuddled him. She kissed him repeatedly. She cooed like a mating pigeon.

Amy beamed with the enthusiasm of her adventures, completely oblivious to the tensions between Jean and Tim and us. This is fortunate, I thought. Perhaps it's best that she be told little for now. Maybe Jean and Tim will be discreet and come out of their inexplicable moodiness.

I'm hopeful that Amy will grab Tim away from Jean, take care of his youthful needs. Maybe she'll also give Jean's insatiable ego a boost that Nancy and I are not feeding as much as she apparently needs. Even if Jean feels she's lost control of Tim and no longer exclusively has his ear, surely she'll see that she's gained some control over Amy's young heart and childlike mind. Yes, Amy's just the kind of motion-machine cheerleader that all of us so badly need.

Tim wasn't with us when Amy arrived. After the initial euphoria of having found our camp and snuggling up to Cole wore off, Amy's disappointment with Tim's absence began to show. I explained that a piece of his luggage and a tent had been lost on arrival, and that he's been making almost daily trips to check on their whereabouts. I said that earlier in the day Tim had gone into Kununurra to wait for arriving buses.

When Tim arrived around midnight, he was happy to see Amy. But he was noticeably distraught. It wasn't that once again he'd had no luck. Rather, while waiting at the bus stop, he'd been approached several times by Aborigines. Across the street from the Hotel Kununurra, in the long rectangular green that divides the main road into and out of town, Aborigines are numerous, loud, playful, and often drunk. Tonight they were falling down, shouting and grabbing at Tim's arms. The women were swearing at the top of their lungs and slugging the men. Many of them, Tim said, begged repeatedly for money, for anything he might give them. It was obviously a disturbing experience. He looked shaken, and for the first time since he and Jean arrived, he was unable to hide his deeper feelings.

Later, lying in our tent, Nancy says that she's been anxious beyond words for Amy to arrive. She feels that finally everything will be different. "This was reason enough for bringing her," she comments.

Joking, my mood better than it's been since the arrival of Nancy and Cole, I say that there are more important reasons to be happy about Amy's arrival. She's brought thirty jars of sweet potatoes and peas and other vegetables for Cole, another twenty of fruits, and bottle liners that we can't buy in Kununurra. She's also come with three different kinds of electrical adapters so that I can now use my laptop.

What would we do without Amy?

4
The Gulf Widens

Tim's different, I swear he is. Or am I merely seeing what I have wanted to see since he and Jean arrived? He's alive, vibrant, and seemingly out from under Jean's moody umbrella. He's more like the Tim for whom I had such hopes when we drank together at the Nineteenth Hole that Friday afternoon. Hurrah!

It's becoming a habit. I walk up the dusty road to the ringer bunkhouse to shave and use the shower, and face the mess there that grows like a virus. Half of a broken mirror, dirty soap and empty deodorant cans, crumpled rags in the deep sinks and on the floor, and today a couple of new additions—a lonely white sock and a black wallet. Two twenties and a ten are visible.

I go into the dark toilet and peer down into the bowl in search of frogs. Almost any time of day I find two or three. I reach for them, eager to get a better look. They're much faster than my hands. Yesterday I forgot to check the toilet before sitting down and one of them played a quick roll of the drums on my bottom.

I shower again in darkness. I remind myself for the fourth or fifth time to get a lightbulb on the next trip into town. The bracing cold shower over, I step outside and grab my towel and reflexively kick my blue jeans and shirt, which are lying on the cement floor. A meter-and-a-half-long black python slithers away from beneath my jeans, headed for a line of water in the corner, under the laundry sink that runs the length of the long wall in front of me.

More good news to note. Tim is finally asking questions about the research. He's also volunteering information before Nancy or I have to ask what he's found. He's taking the initiative to work on data sheets. Now on days when I go into the field, he shows that he can be keenly alert. He's often the first to see a new blackheart nest, a budgie pair preening on a hidden branch, a bat leaving a hollow that might have been used by a Gouldian or longtail. Nancy and I are delighted with this turnaround.

Jean too seems more energetic and somewhat more enthusiastic. The other day she designed a more efficient data form for the information we're taking on trees in the test plots we've chosen to examine in detail.

We're bumping over the boulder-strewn track on our way up into Mallee Boy when Jean turns to me and says, "I would like to do my own research project on longtails. I've got some ideas."

"Great," I answer.

"It's important to get my name on some research papers. This is a good opportunity."

I assume this has to do with her job at the Illinois Natural History Survey. Her name on any research paper can only increase the likelihood of getting a pay raise and secure her hold on the position she was so fortunate to get. "Nancy will be eager to hear what you have in mind," I say. "As you know, she's generous with putting students on a paper if they take some initiative and make a contribution."

As we drive on, I anxiously wait for Jean to elaborate on what kind of research she wants to initiate. She doesn't, and I recall her reluctance all through the fall and winter of 1990–91 to come to meetings when Nancy wanted to talk about papers on zebra finches, field projects for the Australia-bound students. Jean had no time to read what Nancy and others had written on finches, nor to go to meetings. Try as I might, I couldn't get her involved in reading about the Outback or conservation issues that I thought she might find of interest.

When there's no indication that Jean's about to reveal her research plans, I say that a major concern that Nancy will have about a project will be its compatibility with what we've found so far, and with the short- to medium-term longtail research objectives that she outlined in detail for the benefit of the students before leaving Illinois. I remind Jean that Nancy rarely collects data unless she can imagine how it might elucidate significant aspects of natural history or test a hypothesis. I also note that at the moment Nancy is feeling overwhelmed by all we're trying to do. She's worried that we've taken on more than we can handle. "But talk to Nancy tonight or in the next couple of days," I say. "I'm certain she'll be receptive."

Tim says nothing during and following this conversation with Jean. He seems preoccupied, lost in another world.

We leave the van in a charred wood-strewn clearing by the side of the road and cut a trail through the shoulder-high sorghum into Mallee Boy. Nancy wants to see if we can open a small window on what fires do to breeding activity—or at least mid- to late-season breeding activity, since the season is well along. Mallee Boy is one of the sites Terry and Neils burned before the arrival of Nancy and the students.

Tim and Jean and I agree that one sizable section of Mallee Boy that has a moderate-to-high number of breeding blackhearts and Gouldians this season has been completely burned. But after we more or less define a second study plot, Jean and I find ourselves at odds. She insists that 90 percent of the plot has been burned. I say the percentage is probably closer to 20 or 30. I point out that she's ignoring all the high sorghum in the western and southern portions of the plot.

Jean is insistent, her voice strong. "You're wrong," she says.

I scratch my head and look to Tim to participate, volunteer an opinion. He lowers his head and walks away.

Jean and I take another turn through the plot. When we finish, I turn to Tim, who has rejoined us. "What's your estimate?"

"Maybe 40 percent or so. I don't want to get involved."

"Parts of this area we've come through have been burned several different times," Jean says.

"In fact, all at the same time," I respond. "I was here when Terry and Neils torched this area."

Jean points to small circular or irregular patches, most no larger than a couple of feet in diameter. "See these," she says. "These patches are unburned sorghum." She opines that the area's been burned more than once, and at different times.

I explain that Terry and Neils burned Mallee Boy when the sorghum was still wet. I note that a common approach to setting the grass on fire in this part of Australia is to take a box of matches in hand and walk through the grass starting small fires.

Jean holds her ground. She repeats her claim that Mallee Boy has been burned at several different times, and that the burned area constitutes 90 percent of the plot. I bite my tongue, change the subject, say that we ought to spend some time identifying individual birds.

On the trip back to the hotel, I ask Jean what she has in mind for a research project on the longtails.

"I have two projects I want to do. I want to look at variation in longtail bill color, and I want to look at foliage nests."

I wait for her to elaborate. Does she have some hypotheses? She says nothing. Finally, I remark that her idea for a paper on changes in the color of blackheart beaks from yellow to orange or red as one moves from west to east is one of the principal objectives that Nancy had identified in Champaign after she decided to undertake field research on these finches. Nancy had told me in one of our phone conversations that she had discussed this objective with the students. She'd said that if the sketchy reports in the scientific literature were accurate, beak-color variation in blackhearts opened fascinating questions about mate selection, inbreeding and outbreeding, and population mobility in space. The significance of beak color has long been of more than passing interest to Nancy. She spent much of the four years before coming to Australia delving into its role in mate selection in zebra finches. By now she could say quite a bit about male-versus-female beak color and its evolutionary significance, even the degree to which color is a heritable trait. It had taken her, a doctoral student, and scores of undergraduates thousands of hours to gain insight into this single phenotypic trait. Jean is not conversant with either major or minor findings on beak color that Nancy has made, yet . . .

"What do you have in mind on the beak-color question?" I ask.

She won't say. She'll tell Nancy, she says.

Nat arrived yesterday, two days later than expected. A familiar cluttered mind, now chugging along on two narrow tracks. She was so certain that someone would steal her bags, that she roped and locked them together and, whenever possible, refused to let them out of her sight. She made it sound as if, whenever aloft, she'd traveled in the cargo hold.

Nancy's been waiting anxiously for Jean to approach her about her desire to do an independent project. Since I told her about Jean's interest in beak color and foliage nests, she's been trying to come up with a subproject to carve out of the larger beak-color focus, but so far she's been unsuccessful. It's just too early in our data-gathering effort to start partitioning this problem. Besides, Jean thus far has shown no interest in learning how to describe longtail beak color using the Munsell system that Nancy uses as if it were her native language. Perplexed that Jean hasn't yet approached her, she says to Jean, "Rich told me that you would like to have a separate project on longtails. What interests you most?"

"I dunno. Maybe foliage nests."

"You mean, why do birds make foliage nests, as well as nests inside holes in eucalypts? What is their adaptive significance? Let's brainstorm about some possibilities. Do you have any favorite ideas?"

"I think suitable cavities are limiting, so birds have to build foliage nests instead. I've found five foliage nests, and I'm monitoring them. So far there are no eggs."

"That's good. Have you seen any birds building these nests or tending them?"

"No."

"Is there any indication a bird is about to lay eggs in the nest? Are they lined with feathers?"

"No."

"Well, if your hypothesis is right, you need to establish that the nests are used for breeding, and that alternative sites are limiting. The second part fits right into data we are collecting. We're in the process of determining if most of the suitable cavities in eucalypts are occupied, so that part of the work will be done for you. The next step is to find more foliage nests, and to develop some alternative hypotheses concerning their function. We can put them on the map of nests we're building to see if there is any interesting spatial pattern."

Jean says nothing, so Nancy rambles on, trying to stimulate discussion. Nancy notes that the foliage nests that she's seen are fairly conspicuous, not well concealed as one would expect if they were to be used for breeding. They could be decoy nests to distract predators from real nests, but they don't seem numerous enough. In fact, they don't seem nearly as numerous as cavity nests, but then no one has yet looked for them exhaustively. They might be practice nests built by young birds. Obviously, if any birds were seen building them, it would be important to get a look at their beaks to see if they were juveniles or subadults. The nests might even be display sites used by courting birds. Do other birds ignore them or tear them apart?

Nancy points out that a useful way to think about the possible functions of foliage nests is to try to reconstruct why blackhearts have evolved to use cavities. It's notable, she believes, that the birds essentially build foliage nests inside the cavity, although the nests they fashion are not sturdy enough to hold eggs without the support of the tree. Are trees suitable for foliage nesting in short supply in this habitat? Do cavity nests offer greater protection from predators? Are there other possible explanations? Only one or two species of eucalypts seem to be used—why? What about other possible enemies? Ants? Bird mites? Could eucalyptus somehow offer protection against those relentless pests?

Nancy suggests Jean take time to write down some of the possibilities, think about those that could be studied in the brief time she's with us. Also, of course, she should take time to look for more foliage nests on Gouldian Mountain and elsewhere. Jean says she'll do so.

This morning, a bright, clear, warm morning on which Nancy and I woke full of good cheer and love for all, we decided to have a picnic and do some trapping at the reedy water hole where Wes, Newry's head ringer, shot his last cow. Last night, after listening to Nat—seemingly frightened to death—tell everyone about a harmless spider she found shortly after she got out of her sleeping bag, we asked her to join us. She embraced the idea, only to reveal still more fears, including those concerning cholera and intestinal ailments she was sure she'd gotten or would soon get because of the bad water she'd been drinking since her arrival in Australia. Jean, I fear, has fueled her imagination, telling her where our water comes from and that it's full of deadly cockatoo excrement.

The spring, several kilometers south and west of our camp, flows throughout the Dry. For thousands of years it was probably of considerable importance to pre-European Aborigines. The long skinny pool, about half the length of an American football field, is lined with tall reeds and well-developed pandanus and river red gums, which made it ideal for native campsites. The spring has enough water to have supported several groups of gathering and hunting Aborigines for months at a time. Not only did the spring supply indispensable water needs, but because water draws in kangaroos, dingoes, and birds, especially at the end of the Dry when water is scarce elsewhere, the spring must have been a favored hunting ground. Now the spring is pumped to feed the water needs of cattle, who camp here regularly and denude the area of anything resembling grass.

We caught fewer birds than we'd hoped to, and nothing novel. A wind came up, and putting up the mistnets was harder than usual. The big surprise to Nancy and me was that Nat had no idea how to set them up. In Champaign she'd told us that she had all kinds of experience trapping birds.

Cole has developed thrush on his tongue, and as soon as Nancy brings it to my attention we drive into the hospital in Kununurra. The nurse says it is nothing to worry about. She doesn't know whether it's the water we're drinking or something else. Nancy had been boiling water for him. We decide that we'll keep Cole on bottled water for the rest of our stay at Newry. The good news is that he's gaining weight and looks healthy and full of energy. In fact, except for the thrush, he's never looked better.

I use the trip to town to have a tire with a slow leak mended, and to find something to patch the hole in the jerrican—a hole whose origin is a complete mystery. There's no shortage of little things to do.

Jean is getting lots of bites on both legs. The rest of us so far have few or none. She's not the only one wearing shorts, so this doesn't seem to be the reason. Nor is her tent in high grass, or near water, or located in an unusual place. My guess is that they're only chigger bites or something similar and it won't be long before the rest of us get our share. To relieve Jean's anxiety I tell her about the time my father came to Texas when he was in his sixties and we went for a hike through tall grass. The next day I spent an afternoon covering his body with red nail polish so he wouldn't go mad scratching the sixty or seventy chigger bites he'd gotten. He looked so awful he couldn't bear to look at himself in a mirror for weeks.

After dinner Jean and Nat ask Nancy where we are going to send Cole to school. Nancy says that he is much too young for us to have given much thought to the matter. Then, for no obvious reason, the two of them begin lecturing us about how anyone who won't allow their children to be taught by homosexuals is despicably homophobic. The tone is all very accusatory, as if they know that Nancy and I are hardcore homophobes. This strikes me as perversely comical, since had I been asked I probably would have said—and Nancy would have concurred—that we wouldn't care if Cole was taught by a gay alien in pink frilly bloomers and flowing purple hair as long as the alien did an effective job at teaching reading and writing and arithmetic, and made no effort to influence our son's dress or sexual habits. I look over at Nancy and grin; she rolls her eyes and throws me a glance of knowing astonishment, saying in effect, What on earth is this all about—anger over some gay bashing they overhead on their last trip to Kununurra?

Nat is waxing poetic about Dale, an Aussie scuba diver she met on the flight into Darwin from Cairns. She was so smitten that she would've "gotten it on" with him on the plane, but didn't have the nerve to do it in the bathroom and couldn't imagine getting away with it in his seat. Amy, speaking like a veteran student of odd and interesting places to have sex, comments that there are all kinds of ways of having exciting sex in an airplane, and why didn't Nat just use her imagination. Nat flashes Amy a nasty look. There's definitely some competition going on between the two of them, and I doubt that waif Nat can lock horns with worldlier Amy for very long.

Amy seems ideal for Cole and the primitive environment at Newry. She doesn't seem at all bothered by the voracious marauding ants and the dirt and the lack of air-conditioning or other taken-for-granted amenities. Thus far, Amy has proven to be the most adaptable of the lot. And how she loves our boy! She calls him "honey," and she hugs him generously and kisses him on the lips whenever she greets him. She seems to enjoy it every bit as much as Cole does when he crawls around wet and naked, splashing water and pushing his toys all over the gray cement floor in the middle of a breezeless afternoon. On mornings when we leave camp before the sun is up and Cole is still asleep, Amy comes down to our tent and crawls inside our large swag to sleep beside him.

Amy's had no practice with younger siblings as a child, and apparently little or no baby-sitting experience. So Nancy has had to instruct her in most things, even how to change a diaper. But Amy's been open and quick to learn. She's not afraid to ask when she doesn't know what to do. Nor does she seem to take umbrage when Nancy or I correct her or help her with something she doesn't understand.

Still, there are two issues that we feel it necessary to bring to Amy's attention constantly. One is our fear that despite Amy's sweeping and mopping the hotel floor once or twice a day, Cole will find something that one of us has dropped and will put it in his mouth and choke. The other is our desire to keep his skin from burning. From the time Amy arrived, she's taken Cole for daily walks in his backpack carrier. They wander up to the station house to talk with the cook or go over to the cattle yards to watch the ringers sort and brand and dehorn. Sometimes they meander down the dirt and gravel track until they reach the white station gate, pausing here and there to point to and talk at scampering wallabies and the flocking corellas that so fascinate Cole. We're insistent with Amy that Cole go nowhere without ample amounts of sunblock on all exposed skin. We also demand that he always wear the red French Foreign Legion cap that Nancy fashioned for him before leaving Illinois.

A lot of our fear—maybe all of it—has to do with Amy's offhand attitude toward this unforgiving sun. Although I've told all the students that the glowing beast high in the sky is our single greatest enemy, only Jean consistently uses sunblock. Amy seems to positively enjoy a daily toasting. When Cole takes a nap, she heads for a fully exposed hammock, clad only in shorts and a skimpy halter top. And for all I know, she's as naked as a willie-wagtail when we're not around. When she's in the sun, time is never of the essence to Amy—and it shows. Even Tim, not one to carp about much of anything, and rarely with advice of any sort for Amy (at least in public), frequently tells her to put zinc oxide on her ever-red nose. I've come to think that she sees his words as an odd show of affection that can be laughed at and then ignored.

Nat's been pestering me the last couple of days to buy her a couple of rolls of wide duct tape, to seal her tent even more securely than it already is. When she arrived, she said she wanted to sleep by herself. Because we still haven't recovered the tent that was lost with Tim's luggage, I've given Nat one that we thought we'd use for data gathering on days when the flies were unbearable. The tent's roomy, big enough for four people, and it has large breeze flaps on all sides. After I helped Nat put it up at a location of her choosing, she said she couldn't sleep on the ground, that she was afraid of snakes and all kinds of insects that she was certain would do her great harm. In the former cook's quarters in the hotel, I found an old bedspring with hinged legs, on top of which she could put her sleeping bag. But this still wasn't enough protection, she insisted. So I had to buy her a thick and expensive canvas to cover the ground, and then she used all the duct tape we had on hand to tape the edges of the canvas to the bottom of the tent. Nat's fear of being bitten and subjected to a painfully slow and agonizing death by a species unknown to the scientific world is simply fanatical. It's impossible for me to reconcile this with her alleged love for snakes.

I couldn't wait to get home to talk with Nancy about the hypnotizing budgies I'd seen. But she was a step ahead of me, and just as excited—nay, more so. She'd been out in the morning and also registered their fanciful, almost fictitious, flight patterns. Rushing through her mind now was the thought that their activity might benefit blackhearts, perhaps move their breeding into high gear. The brilliantly green little squadron fliers might draw the attention of birds of prey, compound predator choices at mealtime, make it easier for blackhearts to move about less cautiously when nest building, bringing food to young, pairing up, or searching for a home in which to raise a new clutch. Who knows how budgies fit into this complex ecosystem? But who wouldn't like to guess, and then find defensible ways to test the guesses? This, after all, is where science gets creative and rubs shoulders with art.

Snooping about in the farming lands around Kununurra, I learn that in the 1960s this frontier land of once infinite promise was an insecticide hell (Drewe 1990;

Pratchett 1990). Farmers were spraying up to fifty times a season. They were drowning rather than poisoning the *Heliothis armigera,* a ravenous little creature that loved the cotton they were growing. To rid themselves of *Heliothis,* which proved easy in theory and virtually impossible in practice, the farmers killed everything—snakes, goannas, birds, insects. In their hell-bent eagerness to find economic success in yet another tropical insect heaven, they were, it would seem, even willing to risk killing themselves and their families.

Because of spray drift, the frequent aerial application of DDT led to contamination of nearby irrigated pastures. The residues then showed up in the renal fat of cattle grazing this land. There was so much spraying that cattle were found to contain up to two hundred parts per million of DDT. The permissible level for cattle coming into the United States was then seven parts per million. The DDT was still detectable in the fat of cattle grazed on these pastures fifteen years after spraying ceased. The levels of DDT in cattle around Kununurra didn't get down to acceptable levels until 1979.

Lately Amy has been seizing every opportunity to go into Kununurra with me. I'm not sure if it's to get a break from Cole, or to be the first to know whether she's gotten any mail—and find out what the other team members have gotten from friends back home. She's every bit as interested in their mail as her own. She registers return addresses, holds envelopes up to the light, and were she able to get away with it she'd probably open their letters for a quick peek before they saw them.

She's a nonstop question machine. How much did the van cost? When did Nancy and I get married? How long did we live together? How much did we pay for our house? Why are we moving to Irvine? Why do I help Nancy with this research? How many children would we want if we could have more?

Because of much fanfare about rape on the University of Illinois campus before we left, I raise the issue of rape in Greek houses, note that I've read that this is where the vast majority of reported cases occur. I ask if this is a major issue with any of her sorority sisters, which I presume she'd know about since she's been president of Chi Omega. The conversation has a surprisingly short life. It gets derailed when she remarks, "I can do anything I want with a guy and then say no. That's how my sisters feel too."

Taken aback, I say, "You don't see that as teasing, as leading a guy on?"

"No, why?"

"Sounds to me like anyone with that attitude is inviting trouble."

She shrugs her shoulders and frowns. "Rape is no big deal."

Not one who has difficulty with words or swallowing hard realities, I find myself speechless.

I'm finishing a piece of chicken at a small restaurant in town when he walks up to me and says, "What're you doing, boy?" He's tall and skinny, Aboriginal black, with a white beard and an enviable head of white hair, a jagged scar under his left eye that looks like something he might have gotten in a knife fight.

I tell him to sit down, offer him a piece of my chicken and tell him to take all the french fries he wants. He eats as if he hasn't tasted food in a long time—a very long time, I think as he chews on the chicken, then the bones. I gaze at his meatless bare chest, at arms not much bigger around than Cole's.

I stare at a large sore on his lip, wonder how long it's been untreated. I notice that the inside of his mouth is the color of a gardenia. At first I guess that he has eight or ten teeth, then I revise my estimate down to five.

He doesn't want to talk, he's too busy eating. I get up to leave and he says, "Hey, man, give me a dollar." My mouth curls like a pretzel. I glare at him and think of how I respond when bums hustle me at home. One moment I'm angry because I hate beggars, the next I'm sympathetic and giving, because more than once I've thought that I was a centimeter or two from a life of penury.

I walk outside and notice that the trees are spring green and full of life. I feel a light breeze on my face and take a deep breath to smell the clean air. I turn back and look through the window and have a sudden urge to know his age. I think of turn-of-the-century photos of these men, people right out of the bush. All those famous ethnographic black-and-whites that skew my sense of time and human origins. He's just as gaunt, the same beard, the same hair, the same distant stare. Now that stare is more distant than ever.

Nancy has been taking more time recently to watch the longtails arrive at Tanya and Chloe. She'd noticed that they often come and leave in twos; they greet each other with a distinctive head twist and twitter, sometimes sing, and generally behave in ways that suggest they're mated. She's trying to determine if they really are pairs. If so, it will be a great source of information, since many of these birds have been banded and descriptions made of their plumage and beak color. Although it would be convenient simply to assume the duos are pairs (an approach that Jean advocates, based on a study of British passerines she remembers reading), Nancy won't reach this conclusion easily. After a week of observations she reports that while the greeting ceremonies she's observed are fascinating, individual birds participate in them with more than one partner. So far she's not been able to determine which duos are stable partners, which are in the early stages of courtship, and which are long-lost kin having a reunion.

How easy it would be to mischaracterize Nancy, as I'm sure more than one biologist or budding student has done. She's never had the least interest in a Life List. Nor can I remember once in our twenty years together her saying that she'd like to spend a Saturday or Sunday morning in a park or preserve identifying and observing whatever's flying that day. Yet when she gets interested in a particular species, her eyes are as acute as any I've encountered. Her sense of pattern, and what that pattern might mean, is enviable. Invariably, she seems to know just the right questions to ask.

Amy is frowning, seemingly disgruntled. When I ask if something is wrong, she says she is bored. Taking care of Cole and writing letters to friends and reading isn't keeping her sufficiently occupied. "I'd like to spend more time with Tim," she says. "In the field is where I'd like to be." I'm puzzled, this expression of discontent coming a little more than two weeks after her arrival. What are we to do for babysitting?

After dinner, and after I've talked to Nancy, I have a chat with Amy. She reiterates what she said about being bored and then comes forth with a new complaint. She and Tim are shocked at the price of food. They don't know how they're going to find enough money to keep them going and still travel later. They'll have to "reconsider," she says.

"Reconsider what?" I ask.

"Just reconsider, that's all."

In the weeks before Amy came, we decided to tell her upon her arrival that we'd pay her for baby-sitting. Rather than simply reimburse her in-country travel expenses and pick up the bills for food when we all ate together, we'd give her $150 a week, more than what we were paying for full-time baby-sitting in Champaign. We had started to feel guilty: although Amy had more or less invited herself onto our team, we'd come to see our arrangement with her as one of exploitation. By paying her, we'd be comfortable asking her to stay with Cole for six or seven hours at a time while we were in the field. Even though in Illinois she had said she'd stay the whole field season, we thought that giving her a reasonable salary would make it more difficult for her to leave us on the spur of the moment. Increasingly we'd come to think that Amy might prove to be of more value to a successful field season than any of the biology students we'd brought. Thus my surprise with this news of her boredom, her desire to do something else, her need to be in the field with Tim all day long.

Signs of boredom were there right away, I suppose, and I should have seen them. As soon as she arrived, she began sleeping a lot: when Cole napped, later for an hour or two during the warm early-afternoon hours, sometimes even before dinner. I should have paid more attention to all the letters and postcards she's been writing. She's mailed many more in her first week than Nancy and I will send in the entire field season.

So now we've started sending Amy out a couple of mornings a week to be with Tim and Nat and Jean, who are finishing a tree survey on Cootamundra. We're trying to give Amy some free time away from Cole and her Newry surroundings. We've resolved that we'll do this as much as possible, that Nancy and I will tend to Cole while Amy's alongside Tim. We've also started taking Amy to the trap sites, bringing along Cole, his toys and food, and his playpen. We're still paying her $150 a week.

We need to avoid conflict. We can't predict what might happen.

Out of the blue, on the return trip from Dingo Creek this morning, just two of us in the van, Nat announces that she has no interest whatsoever in doing the DNA fingerprinting on her return to Illinois. "I will *not* do it," she says. "I want to do something else."

I'm completely taken by surprise. My stomach is suddenly a knot. I squeeze the steering wheel as hard as I can to control my anger and dam the words forming in my head that I dare not utter.

As coolly as possible under the circumstances, I remind Nat of her promise to both Nancy and me, that doing the DNA fingerprinting was the explicit bargain struck for getting into the master's program and coming to Australia. I remind her of exactly where she made the commitment to me, and how insistently clear I'd been. I remind her of the strings Nancy pulled to get her into the graduate program, to get her a degree that Nancy's department doesn't like to give.

"I feel differently now," she says. "You just have to accept that."

"You have an ethical obligation. You don't have a right to do this to Nancy. Or me. You made a contract with us."

Nat doesn't waver. She's made up her mind. Nancy and I must live with her decision, she says.

Later on, Nancy returns from a long walk-and-talk with Nat. Nancy's face is gray; she's downcast and looks as if she has a bad migraine.

"Well?" I say.

"At first she denied having said that she wanted to quit the fingerprinting. She said, 'I would never say anything like that,' and 'I *love* DNA fingerprinting.' When I persisted, noting the absurdity of her claim that you made up such a story, she suddenly became contrite. She said she was only kidding and is sorry you took her seriously. She was really very sheepish and had incredibly poor eye contact."

"What you do think is up?"

"Maybe she's playing some perverse game with us. Maybe she doesn't know what she thinks. Maybe she was testing the waters about breaking her commitment. Maybe she's as terrified of the lab chemicals as she is of the snakes and insects here. Maybe she doubts she can do the work."

"With her multiple problems with reality, it all sounds very postmodern. At least she's contemporary."

Nancy sighs. "We're spending all our energy on people problems, not research. Maybe we should send the students home, salvage what we can of the field season."

"No," I say. "We're not giving up. We can't let any of them ruin this trip for us. They came to us as experienced field-workers and they committed to the work. They're not children, at least technically. We've also got nowhere to turn. We'd be more than a little lucky to get reliable backpackers in Kununurra to work for us. They might work for us for a day or a week and then leave. And who knows how many birds we'd lose at the net training them. Or how they might rip us off. We've invested too much. We've got to work this out, see it through."

We argue for the rest of the day about what to do with Amy and the baby-sitting problem, Nat and the DNA issue, Tim and Jean who seem to be returning to their attitudes on arrival. I go back over the same arguments, remind Nancy of all the work I've put into her research to make this field stint possible. I remind her that I'm committed to seeing her return to Illinois with lots of data and enough background information to get longtails out of Australia and into her new lab.

Holding her nose, Amy comes over to me and says, "There's a real bad smell around here. Do you know what it is?"

"I'll look into it," I say.

"Yeah, there is something awful around here," Nat says. "Somebody needs to do something about it."

"We'll draw straws to see who cleans the outhouse," Tim says.

"That's great with me, long as I don't get the short one," Amy says. "I won't do it, I just won't!"

"That's not fair!" Nat says.

"I don't care what's fair," Amy says. "I just won't clean out that smelly stuff."

I say, "Don't worry, I'll take care of it."

Jean doubts that the outhouse is the problem. In a rare moment of altruism, she drops her novel, gets up out of her hammock, and begins walking about, looking for the source of the problem. Five minutes later she returns. She's found nothing. She gets into her hammock and goes back to her mystery.

One whiff at the outhouse and I'm convinced that all the meat we're eating isn't the problem. I begin a nose-to-ground search in the direction of the fenced-off paddock just north of our tent. The strength of the scent grows until it becomes unbearable. And then I see what I should have noticed sooner. Beneath the multistoried pandanuses and river red gums that line the Keep River is a dead horse lying on his back. A right hind leg is sticking up in the air, like a misshapen baseball bat. The midsection resembles a pressure-inflated oil drum. I get closer and notice that ants and maggots are searching for entry points into the sealed fleshy interior. As I circle the stiff carcass, the hot rotting smell of death makes me want to vomit.

Later, Nancy and I are out for a quiet evening walk with Cole when we see Alan and his daughter leaving the cattle yards. I run to catch up with them, to ask about the horse. Alan says that a few hours earlier one of his ringers brought the horse to his attention. Until then he hadn't been aware that the old stud he'd recently put out to pasture had died. "I'll get one of the ringers to burn it tomorrow," he says.

He leaves, and Nancy says, "I sure hope they know what they're doing."

She's concerned about all the dry grass around our tent, grass that runs continuously to and beyond the horse. She's aware of the nonchalant attitude of Territorians in the north toward setting grass fires. I'd told her about those Heidi had set the day we were looking at Gouldian finch nests at her study site at Edith Falls, north of Katherine. While walking around, a reporter from Darwin had found us. Heidi, always eager to advertise her Gouldians, went off with the reporter to take photos of them nesting. Presently the two of them were strolling through a field of two-meter-high, very dry sorghum grass. Heidi began throwing lighted matches. It was afternoon, and winds were gusting through the area. Had the wind shifted in the direction of the fires Heidi was starting, she and the reporter would have been fried in a matter of minutes.

After dinner I consider paying Alan a visit to tell him about our concern with a fire around our encampment. But I decide against it. Surely Alan knows the dangers and won't appreciate being reminded by a city-bred Yank. "Alan knows what he's doing," I say to Nancy. "He's probably torched dozens of horses in dry pastures, to say nothing of starting all kinds of fires to protect his cattle herds."

The next morning, Amy, with Cole on her back, is briskly walking toward us as I open the swinging gate into the station. Behind them I can see smoke and the outlines of two station hands carrying water buckets.

"They started the fire a short while ago," Amy says. "The wind came up right after they got it going. It has already gone through the wire fence and is coming toward our tents."

Nancy, now out of the van, is visibly alarmed. I tell everyone to get in quickly. I speed the hundred or so meters toward the hotel and our tents, hit the brakes, and jump out. But before I get three meters, I see that the grass fire has run its course. Only two five-meter logs are still burning. The flames are small. There is, I judge, little to worry about.

Sandy, the peachy-faced teenage jackaroo, and a wiry ringer not much older have large buckets of water and are wetting down the perimeter of the burned area. Not convinced that they've done an adequate job, Nancy gets some buckets from the hotel, fills them, and begins retracing their footsteps. I tell her to relax. She ignores me. I go to the hotel to get something to drink, play with Cole, make some lunch.

Hours later, a large plume of black smoke is rising above the river. The stench of burning rubber fills my nostrils. After Sandy burned the grass for fifteen meters or so around the horse, he piled four tires and several pieces of wood on top of the carcass and then doused everything with diesel oil before setting it on fire. The pyre rages orange and black. Certain there's no danger to our encampment, I catch a ride in a large flatbed truck with four of the station hands. They're going out to a nearby bore to "do a killer," shoot a cow to fill up the station cold room with fresh beef for the next ten days or so, until another killer is needed to feed the crew of eight stockmen and the two cooks their thrice-daily staple of lean red meat.

Wes, the most experienced and knowledgeable of the lot about picking out and cutting up killers, isn't in top form. It takes three shots with his .222 at close range to bring down and finish off a four-hundred-kilo, three-year-old cow. He's effusively apologetic. His manhood and his status as the station's number one ringer are at stake.

Wes and two other ringers take out their twenty-five-centimeter soft-steel American-made knives and get down to the business of peeling away the black hide and cutting the warm carcass into roasts, rounds, and ribs. It takes them less than twenty minutes to complete the job. In all, less than a third of the gross weight of the animal is tossed onto the leaf-strewn bed behind the cab. The hind legs, the bullet-damaged brain, the tongue, most of the organs, and lots of other meat are left where the cow fell.

"Looks like a lot of waste," I say.

"We don't take a lot of the meat because we don't have a mincer," Wes says. "Some of it we don't bother with because Beverley won't cook it." Beverley is the current station cook.

"You mean, she don't know how," one of the ringers says.

There's a long silence. Then Wes says, "Think about all the starving people that meat we're leaving could feed."

There's enough for several meals for a dozen people. Easily enough to feed the grand gathering of dingoes, crows, foxes, and eagles that'll quickly come in and have an all-day and all-night feast. These are savvy scavenging animals who, like Aborigines, understand that in an environment without refrigeration and where fattening feeds come at unpredictable times, you eat and eat until nothing but bone remains.

Dusk is almost upon us as we turn off to the station. The little corellas and sulfur-crested cockatoos are jabbering away, starting to pair up and form into small flocks above and around the stockyards across from the hotel. The horse fire is spent.

I walk over to the carcass. The horse, charred and smoldering, has been reduced to half its live size. The taut brown skin on the face is surprisingly intact and unburned, a death mask best forgotten.

Nat returns from the field overjoyed. She's just seen blackhearts "doing it" in a dead tree. By her account, it took only about two seconds. And that, she declares, isn't nearly enough time for either of them to have an orgasm.

She mumbles something about Dale and what she now realizes she could have done at thirty thousand feet. She adds that the pleasure definitely would have lasted longer than two seconds. When she again finds herself in Darwin, she's going to leave notes for him at places frequented by backpackers. Who knows? Maybe he'll pass through Darwin at just the right moment, and then he and Nat can do what they should have been doing shortly after meeting.

Amy has settled into a fairly predictable eating pattern: a large can of beets, a bowl of ice cream, followed by lots of candy. This is what she eats for lunch, even some nights for dinner. "Saving money," she says. Though why? I wonder. The money Nancy's paying Tim and Amy is enough for the two of them, and with plenty left over. Perhaps they're already looking ahead to their next venture, this time in Asia.

Jean was in town the other day and bought a secondhand guitar for $A80. At dinner she said that this wasn't the extravagance it seems. She's certain that before leaving she'll be able to sell it for at least what she paid for it, probably more.

Nat jumped in and opined that she too was thinking about making this trip profitable. She's going to return home with several cases of Nutella, a chocolate and hazelnut spread that sells in Kununurra for about $A3 a jar. She imagines she can make at least a 200 percent profit on her entrepreneurial venture. Nat eats Nutella like I drink water when running hard at midday.

I sit down at the bowed Formica table in the Rec Room, across from Nancy. She's engrossed in Nat and Jean and Tim's field notebooks. She has asked them to keep detailed notes on their nest searches and to use the ethogram she's prepared for observing longtail nest activity. There are now enough data in their books to decide how effective a job they're doing and whether the information will be of much value.

She asks me to look at the notebooks and give her my assessment. I start with Tim and take notes. "Whenever he sees more than a couple of birds, he writes 'Lots,'" I say. "Tell him to use estimates, numbers."

I read out loud several lines of his hard-to-read handwriting. It appears that he's been watching some interesting nest activity. But after five minutes at a nest he gets up and moves on to a new site.

Nat's notes are all detail, down to exact compass angles on birds coming to and leaving a nest. They're also full of questionable facts and apparent exaggerations.

Nancy shakes her head. "What am I going to do?"

I read through Jean's notes; they seem more promising than either Nat's or Tim's. But none of them strikes me as interesting; all are barely worth the effort. I say, "Unless you can retrain them in short order, I think these exercises are a waste of time."

The students, we agree, need a lot more direction. Yet we don't want to stifle them so that they'll never note the odd behavior, be afraid to write down what might prove to be a novel insight.

Nancy sets about making up more detailed data sheets, one for nest searches and another for nest observations. She spends most of a day designing and then reworking them after getting feedback from me.

We banded 90 longtails yesterday, 119 today. Yet for a run of several days before that, few—a fraction of these numbers. Such oscillations! What do they mean? That all the local populations were at a conference down south, then returned home and were dehydrated from drinking too much? They're getting ready to move on—to the north or east, sensing that already water holes are drying up. If they're to get in another brood before the Wet is upon us and not lose ground in this ongoing, rarely slowing Darwinian race, they'd better find a site where they can stay for a while. Or maybe they had a collective powwow and decided that—stupid humans that we are—if we caught nothing for three or four days running, we'd pack up and go home!

More likely, these oscillations in trapping mean little, except to people like me who insist on an explanation for everything. What an incredible, perhaps rabid, adaptationist I'd make, the surest recipe for an intellectual divorce from Nancy!

I'm now into an afternoon pattern of taking Cole over to the nearby cattle yards—for my own enjoyment as much as his, I must admit. I put his elfin feet on one of the high, thick iron railings and hold him by his toy-sized bottom so that he can peer over and gaze at the dumb-eyed saggy critters. He's invariably mesmerized, by what I have no idea. When I take him into my arms and turn away from the yards, he cries and begs for more staring time.

When enough is enough, I put him on my shoulders and hold him by his forearms. We head off into the open, nearly treeless pasture to the west of the yards. There we often come upon a couple of perfectly still wallabies, lying down or sitting on their haunches. They peer at us, as if too frightened to flee, as fascinated by us as we are by them. As soon as Cole sees them, he bursts out laughing. He giggles and points, squirms and tries to pull away from my grip. Invariably one of the all-ears wallabies turns and darts toward the creek or the sea of eucalypts. The others quickly follow. Cole laughs harder, and he bounces up and down. His laughter is infectious.

Nancy says that she and Nat were watching birds come into one of the water containers we'd put out, when a kookaburra came down from a nearby tree and grabbed something from near the water. As it flew away, Nat shouted, "Look, look, the kookaburra's got a snake!" There was no doubt in Nat's mind about what she'd seen. Nancy followed the kookaburra and got close enough to see with her binoculars. The "snake's" legs were sticking out from the kookaburra's beak.

Nat should write fiction. She is my fiction, friends will say when I describe her fantastic stories to them.

Cheery and harried and late as usual—about an hour and a half late this time—Heidi comes to the hotel to help Nancy identify several grasses. She's brought along a pile of computer printouts on longtail captures and recaptures. The data go back to 1988. Heidi has done nothing so far with the raw data. Nancy is eager to categorize and manipulate them, to bring precision to Heidi's verbal generalities about the finches.

We ask Heidi if she has any objection to our capturing blackhearts at the nest. She says yes, she does. We can't get her to explain her reluctance. But then, a short while later and without explanation, she changes her mind.

"Nat was excellent," Nancy says. "She sneaked up with the net and we got two of the three birds we wanted. We just missed the third one and would have gotten it if we'd been more aggressive."

Nancy and Nat spent the morning in Cootamundra trying to capture blackhearts at the nest. They used a small makeshift net with a large pocket that could be placed over the opening of a tree hollow. As the bird exits, it gets caught in the net. Then it's a simple matter of taking it in hand, drawing blood, and getting data on a familiar list of measurements.

As Nancy recognizes, there's a small risk that the longtails just taken at the nest will abandon their eggs. She'd prefer to catch them when they're sitting with hatchlings; then there'd be less likelihood of nest abandonment. But at the moment we've found no satisfactory nests with hatchlings, and the earlier Nancy gets a firm answer on the efficacy of the trapping method and the likelihood of abandonment, the faster she can turn to other methods if necessary.

There's still plenty of time to get blood from twenty or so families of blackhearts—if they continue breeding, and if we don't suddenly get unlucky with an at-nest capture method that seems on the basis of initial trials to work quite well.

It would be ironic if indeed we got this much blood, for finch blood for DNA analysis is precisely what brought us to Australia. By all measures, it would be a giant double whammy if we got the blood from blackhearts and also from zebra finches.

How will Nancy get money to do the DNA fingerprinting on all this blood? And then there's unpredictable, other-worldly Nat. One minute she says she wants to do the DNA fingerprinting, the next she tells Nancy she doesn't. Yesterday, in a rare moment of breaking rank with the other students, Amy confided to Nancy that Nat is determined that she won't do the fingerprinting upon returning to Illinois. No reason given.

I'm bagging and releasing birds at the net when Neils comes up to me and says, "What are those, holes in the net?" I nod, and he says, "How did they get there?"

"I cut them."

"Oh!" he says with great surprise, a look of disbelief on his face. "When I do this, we never cut nets. We always get all the birds out."

I explain that that's my aim too, but if a bird is badly tangled and the nets are full and the sun is hot and there is inadequate help, I go to the scissors in my knife. What's a couple of hundred dollars and some bragging rights in exchange for the rending memory of dead birds?

Nancy says, "Being here is starting to remind me of life in a university dorm, and I hated it. I absolutely detested it."

Lately she looks for every possible excuse to pick up Cole and go with me on my frequent shopping trips to Kununurra. Yesterday she referred to our trip as a date. She didn't even mind that it meant we had to do all the grocery shopping, pick up the mail, and do everyone's laundry.

We remain utterly puzzled at the students' behavior. Our generosity and attempts at understanding seem only to backfire or to deepen a growing silent rift.

I stop at the town's only real hardware store to buy some Ready-Mix cement so that we can make permanent corner markers on the six study plots from which we've all but finished taking numerous kinds of ecological data—kind and number of trees, potential nesting sites and location, active and inactive nests, and so on. But there's no Ready-Mix to be had. The garrulous American expatriate from Montana who frequently helps me says that Ready-Mix is too costly to bring up from Perth. I'll have to buy cement and then mix it with dirt and gravel, which we can get from the dry riverbeds. The cement costs four times what I'd pay in Champaign. It'll be much higher than this when the new astronomic road fees on trucks plying the Territory go into effect in 1993.

Nancy has taken a day off from trapping to examine some of the accumulating data: phenotypic measurements of the respective sizes of left and right body parts. The data consist of multiple measurements of beaks, legs, and wings. Tim practiced taking these measurements with expensive electronic calipers well before we left Illinois. Nancy wants to see if it's worthwhile to continue the effort. She needs to know if the measurements are consistent enough to be reliable. If they are, she can use the data to see what other traits (such as beak color, bib size, and tail length) might correlate with the degree of symmetry in an individual's body parts. If the data are not consistent, she'll abandon the activity.

For both practical and conceptual reasons Nancy is curious about the results of Tim's measurements. Conceptually, the idea of "fluctuating asymmetry" is that the degree of symmetry of body parts may reflect an individual's genetic quality. Most sexually reproducing organisms are diploid, containing two sets of chromosomes —one received from their mother, the other from their father. When an individual possesses two identical copies of a given gene, that individual is said to be homozygous for that gene. But if the two copies are slightly different, the individual is heterozygous.

Some scientists have reported that individuals produced through kin mating or inbreeding have higher left-right asymmetries throughout their bodies than individuals that are outbred. The explanation for this pattern has two aspects. First, because relatives are likely to have identical copies of genes, inbred individuals are generally more homozygous at many genes throughout the genome (the collective set of genes in an individual) and outbred individuals are more heterozygous. Second, heterozygosity is thought to provide a developmental buffer that enables embryos to develop symmetrically. Scientists have shown that subjecting developing embryos to environmental stresses increases asymmetry. By extension, they have argued that highly heterozygous individuals have greater buffers, which allow them to develop symmetrically.

Originally the concept of fluctuating asymmetry, or FA as it is sometimes known, was developed on the basis of findings for insects and other creatures whose development is not buffered by parental care. Eggs deposited in nature, and subject to vicissitudes of extremes of temperature and of variation in the quality of food available to larvae, may result in large developmental asymmetries. Nevertheless, a number of behavioral ecologists who study birds and mammals have embraced the concept. They have even argued that birds and other warm-blooded organisms (including humans) evaluate potential mates on the basis of their asymmetries. In the jargon of science, fluctuating asymmetry may be an "indicator trait" that accurately reveals an individual's genetic quality, and hence his or her suitability as a potential mate.

Nancy sees numerous problems with this idea. Some are matters of logic. One difficulty is that if fluctuating asymmetry is caused by low heterozygosity, then it's not a heritable trait. "Heterozygosity" is lost when gametes divide and become haploid, having only one set of genes. Gametes must go through the process of ploidy reduction (called meiosis), so that fertilized eggs will again be diploid. If gametes remained diploid, then the fertilized egg would have four sets of genes, a condition that is lethal to most animal embryos. In turn, if a trait is not heritable, then it cannot possibly be an indicator of genetic quality.

Even if the effect were real and heritable, Nancy is uncertain that such asymmetries would be a useful criterion of mate choice. Imagine, for example, a baby bird that contracts a bacterial infection shortly after it is hatched. The infection is a serious stress that impacts development and causes the baby to be somewhat more asymmetrical than it would have been had it not gotten the infection (or had gotten it later in development). But the bird survives. When it becomes an adult, should it be less desirable as a mate because it is slightly asymmetrical, or more desirable because it survived an early trauma? Given the huge variety of developmental traumas that might alter symmetry to varying degrees, it is difficult to infer much from an adult's symmetry unless we make important assumptions, such as the possibility that all individuals suffer similar developmental traumas. If all baby birds contract a bacterial infection early in development, for example, then the degree of asymmetry this trauma produces would likely be more meaningful.

Because so many of Nancy's colleagues embrace the concept of fluctuating asymmetry, Nancy has felt obliged to give it serious attention. Long before coming to Oz, she took a number of measurements on her breeding zebra finches in colonies in Illinois. In the process, she discovered that the degree of asymmetry of most traits was small relative to her ability to measure those traits on live, squirming birds. These asymmetries were so small, in fact, that she seriously doubted that birds could perceive them. She wondered if some other trait, such as beak color, reflected an individual's overall fluctuating asymmetry. But she was unable to find any such trait.

She thought perhaps these results occurred because the birds were breeding in a protected environment in which the developmental challenges that cause asymmetry were minimal. She wondered if her manual dexterity was too poor to achieve high accuracy in taking measurements. She decided that she—or rather, a dexterous student—should measure asymmetries of free-living birds in Australia. She reasoned that one person should make all the measurements to avoid confounding the problem with interobserver differences. Early on, she explained this problem to Tim. He was eager to take up the challenge.

How will she determine whether Tim's accuracy level is high enough? The first step is to identify and compile data for adults that she is sure belong to the same sex and that have been measured twice at Newry. She has found twenty-six such birds that she's fairly certain are adult males. Unfortunately, she doesn't have a computer handy to do a full-scale analysis. But she's able to run simple correlation analyses on a calculator to determine whether the relative asymmetry of various body parts has stayed more or less constant. For example, she subtracts the left tarsus length from the right for each bird, for both the first time Tim measured it and the second. (The tarsus is the long bone in a bird's leg to which color bands are applied; as the word implies, the bone is actually an elongated bone of the foot, analogous but not homologous to the long bones in the legs of humans.) A relatively large remainder from this subtraction indicates greater asymmetry; a negative remainder indicates that the left tarsus is longer. By running a correlation analysis between the first and second measurements for all twenty-six birds, she should get a strong and highly significant positive correlation, because birds' skeletons do not grow in adulthood.

If the analysis is not significant, it means that Tim's error in making measurements is large relative to the actual asymmetry. He's making measurements to one one-hundredth of a millimeter—high precision for field ecology. The all-important question is: how accurate and replicable are his measurements? Small variations in the positioning of the calipers on these small body parts produce a large measurement error. A lesson hard learned by many students is that precision counts for nothing in the absence of accuracy.

The results of Nancy's day's work are mixed. Some of the eight traits show higher correlations than others. Predictably, the larger traits, like tarsus length, show better results than smaller traits, like the length of the back toe bone, or hallux. This is because absolute measurement error is more-or-less constant, so the relative measurement error is greater on smaller traits. She also finds little correlation between asymmetries of the various traits, even the ones that seem most repeatable. This interesting result strikes her as contradicting the logic of the fluctuating asymmetry concept. She thinks she can spot errors in Tim's data, even whole days when he has done a poor job.

Tim is meticulous in this data-collection effort, and it's slow work. Only a few birds can be measured on any given day. The procedure is stressful to the birds, which are held for many minutes in a hot sweaty human hand. Therefore, an individual bird can be measured only once on a given day. For this reason Nancy now decides to concentrate on collecting data from a relatively small set of birds. The new goal will be to take five sets of measurements from each bird, so that "outlier" data points can be dropped prior to analysis. The trick will be to catch enough of these birds. She scrutinizes the banding record to find likely candidates, birds she thinks she can reliably catch again. Then she double-checks their various phenotypic traits to make sure they show sufficient range of variation. She doesn't want to end up with a bunch of short-tailed, large-bibbed birds in this cohort; she wants lots of variation, so she can ultimately determine, for example, if birds with long tails have lower fluctuating asymmetry than those with short tails.

This much accomplished, she makes two copies of a list of these birds, one for Tim and one for herself. They will have to watch for the birds carefully and not miss an opportunity to measure them if and when they're caught. Forty males and thirty-two females are on the list. How many of them will we catch four times or five? Only time will tell. She'll make a note to check on progress of the list weekly, and to watch for birds that should be added to the list.

This evening she explains to Tim what she has done and what she found—and the rationale for her decision, which she reached reluctantly. The birds that fall in the net often aren't likely to be an entirely random subset of the population, but she thinks they have no choice but to concentrate on them. She assures Tim that he's doing a fine job, that whatever the outcome, the effort is worthwhile.

Jean is invariably the first person out of the van when we return from the field at lunchtime. She heads straight for the refrigerator and the leftovers from the previous night's meal. Even more so than with Tim and Amy, money continues to be very much on Jean's mind. She's earnest about saving whatever she can by eating the leftover food that we all buy together. Graduate students and those just out of school usually are poor, but Jean is not the norm. When Eddie died, an insurance policy paid off the entire mortgage on the home they owned.

Nancy has become almost as irritated as I have at Jean's stinginess. Although both Jean and Nat still seem worried about the quality of our drinking water, they steadfastly refuse to boil it themselves or purchase bottled drinking water. Twice recently, Jean has asked Nancy to get some when she goes to Kununurra. Nancy gets the implication: Jean wants us to pay for her bottled water, which we won't do. Despite Jean's worries, she can't bring herself to pay an Aussie dollar for four liters of purified water. The money, it seems, is better spent on a bottle of Victoria Bitter beer, or pocketed as savings.

Recently, Nat has begun to imitate Jean. Today there was a footrace between the two of them to the refrigerator. Nat won, and she promptly filled her plate with enough food for three people. Also in imitation of Jean, Nat made no effort to share the leftovers.

Nat, however, doesn't have the other students' preoccupation with saving money. She'd much rather be eating junk food than leftovers. When she accompanies me on shopping trips, she buys enormous amounts of cheese, vinegar chips, chocolate candy bars, Nutella, and the like.

At times Nat's sense of detail, and her memory, boggle the mind. The other day I asked if she'd seen my knife lying around. She said, "Yes, it's between the chocolate and banana Museli and the bird-banding tackle box." And yes, there it was.

Amy's having trouble sleeping at night. The obvious explanation is that she's napping too much during the day. But maybe she's not feeling well either, what with all the candy and junk food she insists on eating.

Yesterday she woke up sick and nauseous, and she vomited. I took care of Cole while she slept most of the day. She says she has no idea what's wrong with her, commenting only, "I get sick a lot. I'm a throw-uppy person."

I offer to take her to the clinic at the hospital. She says no, she'll be fine.

We've now caught and banded about 325 longtailed finches and 66 Gouldians. We've also recaptured another 80 longtails that were banded in previous years—1988, 1989, and 1990. Although the overall return rate of 20 percent from previous years is not strikingly high, Nancy finds it reassuring that there are *some* recaptures. The received wisdom is that small songbirds such as these have constant, usually high, mortality rates as adults. If this were true for blackhearts, then we should recapture far fewer birds banded in 1988 than in 1990, because a much larger fraction of the birds banded in 1990 should still be alive. But thus far we've recaptured about the same number from the two years. This might suggest that adult birds are longer lived than commonly expected. On the other hand, the trend could be the result of uneven banding intensity across years. If Heidi banded many more longtails in 1988 than in 1990, the fact that about the same numbers return from each year obviously means less than if she banded with equal intensity. On this point, the historical record is equivocal. Heidi banded somewhat more birds in 1988, but perhaps not enough more to account for equal numbers of returns. Nancy opines that it doesn't pay to look at this aspect of the data much more closely until the season is over and all the data are in. Still, I've asked her to speculate what might happen between now and the last time we take down the mistnets.

She notes that thus far most of the birds recaptured from previous seasons have entered the nets early in our trapping. This, she thinks, is significant. Because of what we found for zebra finches in Alice Springs, she suspects these birds are residents that breed nearby, and that we'll continue to catch them from time to time during our stay here. We're also catching birds that have been here one or more years but were never banded before. This is the result of Heidi's not trapping with high enough intensity to band more than a fraction of the birds in the area. Unfortunately, once a bird reaches adult age—which Nancy puts at about 135 days for a blackheart—we can't determine its age based on appearance. Thus, apart from the birds that Heidi has banded, we can't identify birds that have resided in this area in previous years.

With our frequent trapping, however, Nancy believes that we'll be able to distinguish "residents"—birds who stick around awhile—from "flythroughs"—those that stop for a drink on their way somewhere else. The differences are already apparent in the data, which is a big surprise to Nancy. Retrospectively, she'll be able to look at the phenotypic data to see if there are quantitative differences between birds classified as residents as opposed to flythroughs, a distinction she was able to make for zebra finches in Alice Springs in 1986.

I'm somewhat puzzled about Nancy's surprise that there appear to be residents and flythroughs. She explains that she never expected to find this difference in zebra finches in Alice Springs. Based on what has been written about central Australian birds, the expectation was that insectivorous birds like butcherbirds and fairywrens would remain territory residents through an extended drought, but that small seed eaters, which include zebra finches, would quickly move on. Finches have no territory to defend, and grass doesn't set seed in a drought. Grass seed is the principal component of a grassfinch's diet. So there would seem to be no point in staying, and every reason for leaving a drought-stricken area. Therefore, Nancy did not expect that some zebra finches would stay put and sit out a drought, while others would be itinerant.

The finches that stayed put during a drought had to rely on the seed bank—old seed that's stored in places like cracks in the dry soil—rather than fresh seed. In this predicament, search time for food would be so high that the birds wouldn't have the time or energy to reproduce. Because a central Australian drought is likely to be long relative to what is thought to be a typical finch lifespan in nature—where ravenous butcherbirds and falcons lurk near every finch watering hole—it doesn't appear to be good strategy for a finch to try to wait out a drought. Perhaps, Nancy reasons, some birds just get stuck; they don't realize they're in a drought until food becomes so limiting that long-range travel is prohibitive. But now Nancy thinks that alternative "lifestyles"—akin to humans (I'd say, but she wouldn't!)—are available to finches. Perhaps, she reasons, individuals that are best at locating hard-to-find food are less likely to join feeding flocks. Or they forage in smaller groups. As the abundance of food declines, those birds that rely on new seed to meet all their energy demands have to travel on to satisfy their daily needs. But climatic conditions are so unpredictable in central Australia that traveling is a risky undertaking. And it's by no means certain whether to head east, west, north, or south. So for individuals that can exploit the food bank most efficiently, staying put may be the best option. Nancy would call it a "life-history tactic."

The ecology of the Top End where we're trapping is very different from Alice Springs, which leads to somewhat different expectations. Simply put, the seasons are more predictable. Here it looks as though blackhearts breed in these savannas in the Dry and go somewhere else in the Wet. There is no compelling reason to expect an itinerant class of birds. But something else comes into play critically with the longtails. They simply can't be as itinerant as zebra finches, because the resulting gene flow would quickly erase the beak-color differences we've been finding among various populations. This has led Nancy to reason that most finches here would prove to be residents. Yet the growing data base suggests this is not the case! It looks as if many finches will be caught only once. Of course, it's possible that some birds are more sedentary than others and perhaps some don't even leave the savanna at the end of the Dry. To get closer to an answer, one at least a bit more satisfying, we'll have to wait and see whether the birds we've recaptured from previous years are still flying into our mistnets when other birds have gone north late in the Dry.

The author releasing finches from a mistnet at Dingo Creek. The birds are held in the two-compartment box shown, then released after they are banded, blood is drawn, and measurements taken.

A brown falcon near our data table, waiting for the opportune moment to grab a finch in flight.

Nancy retrieving birds from a net at a billabong, a water hole in an otherwise dry riverbed.

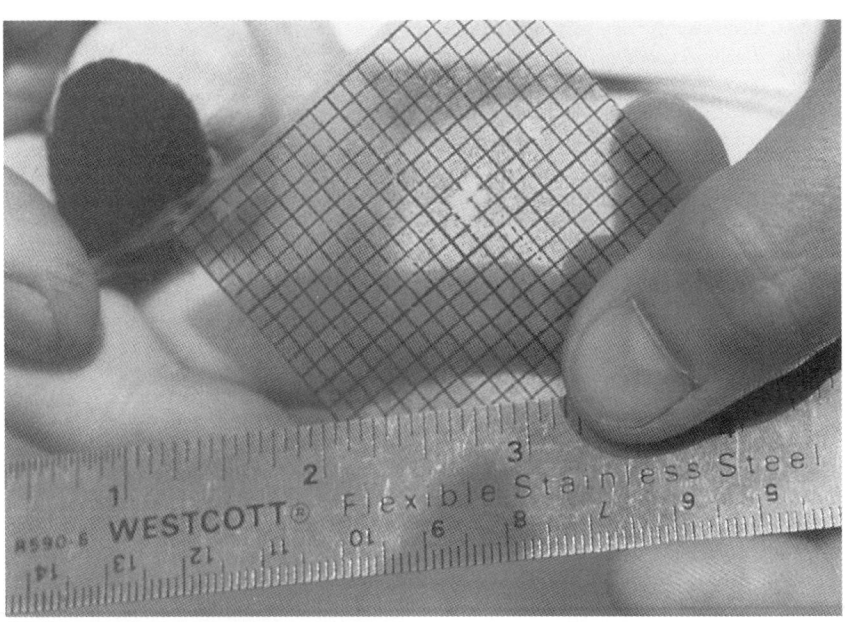

Nancy measuring the size of a longtail's breast patch.

Itinerant backpackers picking melons on a farm in the Ord Valley, north of Kununurra.

Nancy holding a very large clutch of longtails between one and two days old. Contrary to popular myth, holding of such birds by humans does not cause the parents to abandon them.

Cole at a trap site, fascinated by a willie-wagtail caught in our nets.

An azure kingfisher captured in a mistnet on an exploratory trip into Western Australia.

Sorting cattle in the yards at Center Camp, about fifty yards from the Blackheart Hotel.

Terry and some of our team members checking for longtail and Gouldian nests.

Nancy putting seed into a walk-in trap.

Finches are trapped both at the water trough and around the "turkey's nest"—the earthen water tank in front of and to the right of the windmill.

I'm in the Rec Room trying to catch up on my journal when Jean and Tim drive up in the van.

"We're out of water again," Jean says.

"Don't forget to check that the generator's on," I say.

Ten minutes later Jean returns, this time alone. "Still no water," she says.

"Be patient," I say. "The system here's pretty primitive."

"I climbed up the ladder and saw that the tank was empty."

"The cattle get water first."

Fifteen minutes later Jean's back a third time, again alone, and again in the van. Now I remind her that it could take several hours before we get water. "We're last in line at the spigot," I say. "But we'll get water before long. The yards are full of cattle and they're too valuable to forget."

She gives me a puzzled look, and I sense she's not satisfied with my explanation. I say, "I'm confused by your urgency. What's the pressing need?"

"Nancy needs the water for cooking."

"She does?" I ask. The evening meal is at least two hours away and all we're having is shish kebab and some vegetables that I'm going to wrap in aluminum foil and bury in the hot coals. And then I think: this kind of demand is uncharacteristic of Nancy. Patience—far more than I've ever had—is one of her virtues.

Eager to get Jean off my back, I tell her to follow me and we'll do what we can to solve the problem right away.

We find Sandy sitting outside his bunkhouse room. He explains the problem to Jean much as I'd done earlier. She's still not satisfied. I then ask Sandy to do me a favor and take her over to the water valves, explain how they work and in what order everyone gets water. He cheerfully obliges.

Later, after I build a fire in which to cook the onions and potatoes and squash that we're having with T-bone shish kebabs, I ask Nancy why she'd been so anxious to get water. She gives me a look of incomprehension. I tell her about Jean's angst.

"I didn't tell her I needed water," she laughs. "Jean just wants to take a shower before they leave for Keep River." We'd agreed to give them the van so they could spend Sunday hiking in the park. Now it seems they've also made plans for a Saturday night party, and the sooner they can get away the better.

As the students are about to leave, Nancy reminds them to ask Terry and Neils for an update on where they're finding breeding longtails. "We need the information for scheduling the coming week's activities," she says.

We get up around eight the next morning. I build a blazing fire and put on a full billy of coffee grounds and water while Nancy cuts up a couple of potatoes for home fries. After attending to Cole's needs, we spend a happy family hour around the fire, leisurely chatting about our progress. I note that with occasional exceptions the students continue to show little or no curiosity about the research, and that I'm appalled by their methodological and philosophical naiveté about biological research. "Two of them are master's students and one was the outstanding undergraduate in your department," I say. Laughing, I add, "I don't even have a degree in biology. Come to think of it, I've never had a formal course in biology."

Nancy reminds me that it's been several years since I've been around students, and that I have no idea just how good Tim and Jean are compared to most at the University of Illinois. She also reminds me of the age difference. I resolve to lower my sights.

We remark to each other how grand it feels to have this respite from the students. We decide to encourage them to take more of these trips, and also to get away for a night or two on our own in Kununurra when it's our weekend to have the van. When Tim and Jean arrived, I wanted to ensure that the students felt they were being treated as equals, and to give them a chance to see as much of this part of Australia as possible. I'd told them that short of an emergency they could have the van every other weekend.

By late morning Nancy's merrily buzzing along with data analysis, and then the construction of still more ways to sort it, analyze it, reassess where we're going, decide exactly what we should do next. I take the opportunity to play with Cole and bring my journal up to date and catch up on reading the local and national newspapers that have piled up in the previous week.

When the students return just before dusk, Nancy can't wait to find out which trees have new longtail breeding activity. Earlier in the day she'd worked out a schedule for the first three days of the week ahead. The first two days were to involve watching nesting activity at new nests, those that Terry and Neils had found the previous week.

"What did you find out about breeding activity?" Nancy inquires shortly after the students arrive.

"We didn't get the information," Jean replies. She adds that they'd spent Saturday night with Neils and Terry, but no one bothered to ask them for an update on blackheart nesting activity. Jean says they all assumed that they'd get the information on Sunday. "But they were gone when we got up in the morning. And they weren't there when we were ready to come back."

Nancy is obviously perturbed. She'd twice made a point of telling them not to forget the nesting activity information. But as usual she holds her tongue, saying only that in the future she expects one of them to take responsibility and bring back what she asked for. None of the students responds. Presently they leave to take a shower, listen to music, read a novel.

The next morning, Nancy says that we ought to make a special trip to Keep River to get tree numbers on new longtail nests so we can move ahead quickly on getting behavioral data. Before she has a chance to elaborate, I have to tell her that we can't. Jean has just given me their mileage for the trip. They'd put on more than twice as many kilometers as they'd said they would. Where they had gone we weren't informed. I tell Nancy that the gas remaining in the tank and in the jerrican beside the hotel is for Monday and Tuesday's field stints and the Tuesday shopping trip into town.

We resolve that henceforth we'll assign responsibility for everything—even trivial tasks—to a particular individual. Contrary to our expectations before leaving, and in spite of many discussions with the students about being aggressive in the field and not waiting for someone else to do an assigned job, it now seems abundantly clear that they're not willing to shoulder any more responsibility than we demand from them.

This is the second day in a row that we've trapped at Dingo Creek. Everything is different from yesterday. We captured only half as many finches today. It took a solid half hour for the blackheart population to build up to fifty or so birds in the trees near the water hole. Then they'd suddenly flee without apparent reason. No wind, no big flying predator that we could see, no change in behavior on our part.

We spotted lots of brown quail on slow, long marches to the water, and the first gallah we've seen at any of the traps. We got a couple of budgies in the nets, which are uncommon and unwelcome, much as I love to see their brilliant greens and acrobatic flight patterns. They bite hard and occasionally draw blood. Nat was upset that she didn't get to release them. I would have been happy to give her the task if she hadn't been busy. I think she wants some visible battle scars, material for spinning yet another fiction.

We caught a few Gouldians and took blood from them. We're also taking off some of their colored bands, because of swelling on the legs. Heidi says she'd never noticed this problem until Nancy pointed it out to her recently.

We continue to give all the Gouldians we catch Invermectin, Heidi's air-sac mite cure-all. I'm beginning to wonder if the drug might have unknown adverse effects—if it could be killing the very birds she's so determined to save. It may be time-consuming and inconvenient to test the drug, but I for one would like some assurances that the medicine's safe.

Terry showed Tim two bowerbird nests this morning. The walls were made of straw, much of it black. The floors of the nests were covered with white pebbles and glass—the latter probably from beer bottles. Tim beamed at the sight of these two nests, and he wanted to talk about little else the rest of the day. If only he'd show this kind of enthusiasm for our research!

I'm pouring coffee into the boiling water when Nat comes up to me and says, "When are you going to buy a flyswatter? I really can't stand the flies around here."

"Start drinking some of this black pea soup and they'll stay away," I say, trying to humor her.

"No, I'm really serious. We must do something about these flies. They're killing me."

I stir the grounds and tell her there is no fly problem, not even close to what I imagined we'd have to deal with. I remind her of stories I'd told at our Sunday get-togethers in Champaign, about the fly hordes in the spring of 1986 in Alice Springs, when they covered Nancy's face and mine. I tell her how when we trapped south of Keep River National Park with Heidi that year there was so many flies on our faces that from fifty meters we could have easily been mistaken for Aborigines or African Americans.

Nat frowns and groans, and I say, "Try wearing a face net. Nancy brought one for each of you. She's probably even put your name on it."

"Forget it, I *won't* wear one! Terry would make fun of me." She stomps away.

The coffee's boiling. I pour myself a cup, and the delicious smell fills my nostrils. I smell again and blow hard through the steam, thinking, If there's nothing like this in heaven, I'll find another place to spend eternity.

We've finished trapping for the day, and we need groceries and gas. We have to pick up the mail and do the laundry, get the newspapers they're holding for me. Everyone except Nancy wants to go into Kununurra.

We're on the Victoria Highway, a hundred yards outside the station gate, when Jean initiates a game to see who can distinguish the flavor and color of the candy called Skittles, eyes closed. The serious game continues for more than forty minutes, until we reach the outskirts of Kununurra. The consequential conclusion is that Tim's able to make almost no distinctions based on color. Jean, on the other hand, can by taste alone identify all the colors except green and yellow.

Tim and I decided to trap Dingo Creek this morning while Nancy took Nat and Jean over to Chloe to set up nets and trap there. An hour and a half after Tim and I had the nets up, we hadn't seen three blackhearts and there were no Gouldians. When a breeze became a light wind, we rolled up the nets and decided to explore the southern reaches of Dingo Creek. We went looking for water, small flocks, nests, any kind of surprise that might enlighten us on the ecology of longtails. After two hours of walking the banks and the creek bed itself, we had come upon two small but dry depressions and five trees that Terry and Neils had tagged. Without a ladder we couldn't check the nests. We did find one nest that I was sure belonged to a zebra finch, but given how few we've seen since I arrived, it has no meaning for our research.

Amy told me this morning that Jean's been complaining to Tim and Nat that what we're doing is "basic"—by which Jean means simplistic, not nearly as challenging as what we might be doing. As Amy tells it, Jean also opines that we're "going slow," and she's upset because the work is "tedious."

What are we to make of these complaints that come to us only secondhand? Certainly the work is basic: almost nothing is written up in the scientific literature about long-tailed finches. I often joke with Nancy that even with the few data we've gathered and what we've seen, we're already "world experts" on the birds.

We've chosen study sites, we're making behavioral observations so we can identify the sexes, we're determining and assessing sexual and social roles, we're trying to decide when and how we can use zebra-finch behavior as a source of hypotheses about longtail behavior. Many of the ideas emerging in these first weeks may not be tested by Nancy or future Ph.D. students for years to come. Yet once identified, they'll always be there, begging for answers, modification, rejection in the light of hard-won and tediously collected data.

Nancy's perception, and mine, of our progress is precisely the opposite of Jean's. Indeed, we're amazed at how far we've come in these first weeks: how much we've learned, how few false starts we've had, how many tentative hypotheses we've formulated. We frequently feel that it is the students who are slowing us down!

Regarding the charge that the work is tedious, I'm tempted to tell Jean to join Terry and Neils for a spell. Day after day they climb ladders in search of new Gouldian nests, and beyond a little trapping—which only began after we arrived—they do little else. Tim and Jean and Nat's jobs are quite varied. We make certain that each of them gets to do a little bit of everything, and from the beginning we've made morning and afternoon activities, and those at the beginning and end of the week, different. In the time that the three of them have been with us, they've mist-netted birds, searched for new water holes, looked for new nests and breeding pairs of blackhearts, prepared maps on previous nesting activity in 100m × 100m test sites, tagged trees and measured their heights and identified new species, bled and banded and taken a variety of measurements on several species of finches.

What does Jean suggest we should be doing instead? She has no suggestions. None at all. All this, I fear, is just another aspect of her ill-defined discontent. Or is the explanation for her bad-mouthing more insidious, a sort of one-upmanship to keep the other students firmly under her thumb?

Nat's been having a spat with Jean and Amy, telling them not to use perfume and hair spray, which make her sick. I haven't noticed anyone here using perfume or hair spray, and Jean and Amy deny doing so. But Nat insists that somebody's lying; her nostrils tell her so.

Frustrated that she can't find the alleged culprit, Nat reaches in her backpack and pulls out a jar of Nutella, a spoon, and a large package of vinegar potato chips. She then heads for her hammock and a familiar ritual—by the spoonful. I expect Amy to join her in Tim's hammock any minute now. Lately Amy's been eating half a jar at a sitting. How they manage this without gaining weight is one of the small mysteries around here. One of these days I expect them to have a contest to see who can eat the most jars before dinner. As they lie dying, I can rush them to the hospital, call up a local journalist, and tell him to rush to the emergency room, where he can have an exclusive on the first confirmed case of death by Nutella poisoning in Australian history!

Heidi has gotten herself in trouble with her bosses at the Conservation Commission. In a radio interview she announced that she was at odds with the gold-mining company that's paying her for research on Gouldians around Edith Falls. The mining company professes to have found ten Gouldian nests in the area where it's blasting and digging for gold. Heidi claims to have reached at least nineteen active nests, and several more in another area where the company says it found none. She has also said that the Gouldian population in the Edith Falls area ranges between five hundred and eight hundred, a figure far higher than the gold diggers want to hear. How Heidi arrived at this number I have no idea, and I suspect the numbers are more public-relations rhetoric than scientific fact. She's also angry because the company won't pay the money it contractually owes her—not until it sees exactly what she's written about them. Surely she knew that in taking the mining company's money she was being asked to come up with numbers that would minimize negative public reaction against the venture?

Once again it sounds very much as if the Conservation Commission is in league with private industry, exploiting rather than protecting a valuable resource. The commission, Heidi says, gives the mining company access to all data it's collected in preparing its own environmental impact statement, and it also looks over the mining company's statements and tells it what needs to be corrected to pass muster. It's yet another reminder of the nasty politics of conservation biology: that facts and findings are driven by agendas that are not only political but social and economic, even personal.

I'm watching a charming pair of perky wallabies feed in the tan grass across the river. They take a bite or two, straighten out their ears, look around to see that danger's not imminent, and then return to their early breakfast. They seem oblivious to my not-so-distant presence, and to the score of preening and screeching little corellas sitting in the pandanus and gum trees high above them.

Recently Alan told me there are between four hundred and five hundred wallabies within three kilometers of where we camp. Of the three species, the most common is the agile wallaby, the only native animal in the Territory that's not protected. Alan said that wallabies are pests and eat food that foals should have. He ought to shoot them but doesn't, because it wouldn't have any effect on their numbers.

Apparently Jean hasn't learned all she wants to know about human variation in ability to taste nuances in Skittles. She continues to initiate "serious" Skittles discussions and, anxious to get an appropriately large scientific sample, is undertaking more test trials. The Skittles Study spilled over into the field today, and it made Nancy quite irritable. From the beginning she has insisted that everyone at the banding table be resolutely focused on measurements and the bands being put on the birds. One or two absentminded mistakes can create hours of additional work or, worse, go unnoticed.

Nancy said to me, "Jean is either being incredibly mean-spirited in taking this route to let us know just how uninteresting she finds the research, or she is the most frivolous twenty-eight-year-old widow in history."

I walk through the empty cattle yard, climb up onto the third steel bar, and fix my eyes on the black smoke that's pouring out of the tall rusty chimney that sits atop the branding-iron fire. Stuart, the heftiest and friendliest of the four Aboriginal stockmen, is shoving and kicking heifers into the squeeze chute. He's got a family somewhere down near Halls Creek in Western Australia.

Mick, the swaggering silent one with the huge tattooed arms and the bad attitude, is working the chute. When a heifer gets inside just far enough for the head to stick out, Mick slams the tubular iron brace around its neck and midsection and gives the contraption a couple of extra pushes for good measure. Then he throws the unwary young critter down onto an old tire and goes to work, first with an ear clipper to cut away a small hunk of flesh as an extra identity marker. Then with the long-handled pipe cutter he snips off the horns. He goes back for a second cut if the first one's not close enough to the skull.

"There's more pain here than meets the eye," Wes says, coming up behind me, "much more than getting the nuts cut off." He explains that all the heifers at Newry are dehorned to prevent bruising of the meat. Bruising shows up in the abattoir and along with all imperfections in the meat or other usable parts, including the brain, results in a debit to the station account. Dehorning was only started a couple of years ago at Newry. Until now the effort has been pretty haphazard.

Wes leaves and I turn and watch Mick. He kicks the lock on the squeeze chute and opens it, then gives the next yearling in line a hard kick. The processed heifer screams and squirms and rolls out into the dusty yard. Confused, disoriented, it scampers to its feet and slowly moves away to join the others.

Jenny, behind Mick and Stuart in the adjoining yard, is using a charged rod to push cattle back toward Wes and the others, who are vaccinating for botulism. Jenny is only sixteen, not a steel-hard thirty-something male, I remind myself every time I see her. She chain-smokes, she struts about in oversized army fatigues and combat boots, she scowls and growls because the days are too long, and she feels "bloody awful" and wants to quit. Jenny loves to declare that she's quitting, if not in the next hour then tomorrow morning before sunup.

I leave the ringers and Jenny to their business and cross the dusty track and pass by our water supply. I continue on beyond our tent, a burning hothouse at this time of day. With sunset still a couple of hours away, I skid down the embankment and follow the meandering, wet-and-dry riverbed in search of kookaburras. After a half hour, nothing new or captivating in sight, I circle back and head for the hotel. Just beyond the white gate I see Amy and Cole standing on the road across from the cattle yards.

"Cole has just taken five steps!" Amy cries as I get within shouting distance. "He's so good he'll be walking any day now," she says as she takes him up in her arms and gives him a big kiss and hug.

How I love her passion!

For several days I've noticed that whenever I see Amy, she's pushing Cole to walk, goading him on, pulling him along. Yes, and no, I think. I must find a way to ask her to slow down and let Cole find his legs at his own pace. But she's doing a mighty fine job with him, and I don't want to criticize or upset her. We have enough problems already.

At Newry the station hands earn $A230 a week for working twelve-hour days, seven days a week. This is more than they get on stations in Western Australia, about $A35 more than stations in South Australia pay.

The Newry Station ringers are charged $A60 a week for room and board. They get no fresh vegetables, no fresh fruit, eggs rarely, and beef every day of the week—often three times a day. The young trainees, the ones they call jackaroos and jillaroos, get about $A100 a week for the same work, plus some spending money from the government.

The cooks whine about inadequate stores, antique appliances that don't work, the lack of fresh vegetables, the tightness of the budget, the meager quantity and poor quality of range beef they're cooking all the time. There's never any fish or chicken. Chewing on chicken to a Newry horse rider is like savoring filet mignon to someone in the American working class.

But the cook is special, even when not very capable. She's paid more than the head stockman. Only Alan, the station manager, makes more.

The homestead has none of the quiet that the isolation would suggest. Twenty-four hours a day, we are enveloped by the pounding rhythmic thump d'thump of the station generator—power for the lights and the pumps, everything that requires energy. Until you've been at Newry for a week or two, the noise is as disconcerting as living down the street from an all-night Latin dance hall.

The first thing the station hands seek out when they return from their long days of working cattle or mending fences and bores is their allotted beers. Two, and that's all. They gulp them down, then try to relax. They hardly think about anything more serious than flipping through a comic book or a show-all girlie magazine. Some drop their heads into tired hands to watch a little television—imported rerun programs filled with snow and static and little of value. By nine or nine-thirty in the evening, they're all in bed.

Amy was upset on our food run into Kununurra today because her and Tim's portion of the food bill was "too high." She said, "I'll have to talk this over with Tim so we can cut expenses." Apparently, despite the fact that she's now able to save as much as $100 a week, she has a serious money problem that she didn't have before leaving Illinois, before knowing that she'd be paid for baby-sitting.

I wonder if she and Tim have a new and "enlightened" perspective on their postgraduation Australian adventure. Whereas previously it might have been difficult for them to do much traveling in Oz and Southeast Asia after their stint with us was finished, now, with much more money at their disposal, they will be able to spend several months bumming around. How ironic that Nancy and I perhaps have created the very situation we so wanted to avoid—that of making it easy for Amy to say she wants to leave earlier than we'd originally hoped she would.

"What should we have done differently?" I ask Nancy later.

Her face grows long and she shakes her head in bewilderment, an all-too-familiar sight these days. We reassure each other that whatever the outcome, we've done the right thing in paying Amy for taking care of Cole.

Nancy suggests we talk to Amy about our aims, to get assurances that she and Tim won't leave before we've accomplished some minimum goals. I, however, think that it would be better temporarily to ignore Amy's statements. "Let's wait and see what she's saying a week or so from now. We can deal with the issue of their abandoning us when it arises."

Nancy is frustrated with the poor outcome everyone is having in recognizing band-color combinations of birds during field observations. The most obvious way to improve success is to get closer to the birds by using a blind, what Australians call a hide. But were we to make one out of burlap or similar material, it would move in the frequent winds; the birds would be frightened and either leave or behave unnaturally. From previous experience we've noted that it usually takes quite a while for finches to habituate to the presence of a blind. Perhaps we can build one out of papier-mâché, paint it, and make it look like one of the termite mounds that are so common in the Territory.

With materials bought in Kununurra, Nancy and I begin fashioning a frame out of double-strength chicken wire. Then with old copies of the *Australian* and other newspapers that I've accumulated since my arrival, a large bucket of water, and several kilos of flour, we start the messy job that neither of us has attempted since childhood. A little more than halfway through the effort it's apparent that we have miscalculated how much weight the chicken wire can bear. Even when dry, and without the job anywhere near complete, the frame sags in one direction or another, or collapses. Since we cannot get heavier mesh in Kununurra, we abandon the idea.

Cole reacts with a knowing laugh when I point to little corellas and call them cockies. I frequently direct his attention to the swooping, screeching, nerve-grating birds when I take him up in my arms in the early morning or just before sundown. I cannot now remember if he was with me recently when I learned that *cockie* is an Australian term for farmer or station manager, and that among some it is pejorative, meant to identify a redneck, a yokel.

I haven't yet tried to count the cockies here at Newry, and it's just as well. There are too many; the task would be onerous. In any event, they don't stay still long enough for a true and accurate reckoning. They're too scattered, too mobile, too itchy—screaming all the time, flitting up and down and through the branches of the river red gums and the middle-aged pandanuses that crowd the long eastern boundary of our encampment.

Even if all the little corellas at Newry were uniquely color banded, how many students with near-perfect memory of the color combinations would it take to carry out a decent behavioral study? Ten? Twenty? Who knows. . . ?

Nancy and I have numerous insect bites on our legs. I've got about forty, Nancy half that many. They're probably chiggers. We guess that we're getting bitten so much because of the location of our tent, the grass around it, the nearness to the creek below. Or are we getting them in the field while tramping through chest-high sorghum grass?

Tim and Nat have another problem: rashes the size of oranges on their legs. Tim's rash looks ugly, and it's spreading. Whether it's from something they're rubbing against because they wear shorts in the field or, in Tim's case, a reaction to all the mangoes and cashews he's eating (his hypothesis), no one knows.

Tim takes it all in stride, doesn't seem to be concerned.

"If it gets much worse, I think we ought to take you to the hospital," I said to him yesterday.

He shrugged his shoulders and shot me a look of studied indifference—all so familiar.

The last two days we've caught few birds, and I expected the same today. But Nat and I worked the nets for five straight hours, and what a catch! Like nothing I've seen before. More than 155 longtails, and even more than that number of doves, pigeons, honeyeaters, and double-bars that we released. A significant number of the longtails were juveniles. We also got eight Gouldians. We even managed to get three female zebra finches, among the very few we have captured at our three principal trap sites. Nat's fascinated with their bill colors, as no doubt Nancy will be when she sees them.

On the down side, and not surprising given the numbers, two of the birds died. One from a broken neck, and one apparently from the heat—though the day was no different than previous days, and the bird was not long in the net. Perhaps it died of fright? Nat had found the masked finch with the broken neck and had had to kill it. When she came to tell me about it her knees were wobbly and she was crying. She said she'd never had to put a bird to death before. In the lab she always left that task to Nancy. I tried to comfort her as best I could.

It's surprising that we only lost one other bird to a hawk, and not because of the nets. When he made his kill in midair, high above the pond of water, there must have been two hundred finches in the nearby trees. Once the hawk had the finch in claw, he flew no more than thirty meters to the top of a gum tree before perching to enjoy the meal. The hawk ate as if starved. Fewer than twenty minutes after the hawk made his kill, a hundred or so finches returned to the waterside trees, as anxious as ever to get a drink.

Nat did a superb job at the nets, and I complimented her a couple of times. But for reasons I don't understand she takes a compliment poorly. Both times she questioned my sincerity.

Neils came by to get the Gouldians to take them to Terry, who's at Chloe. Together they'll take some measurements before releasing the birds. I asked Neils to take the first batch right away because we'd had them over an hour. But he persisted in dawdling, sitting with his video camera in hand hoping to get some unusual shots of Gouldians. I had to get insistent with him, such was my fear for the birds. Neils thinks and talks of almost nothing these days besides the film on Gouldians he's making.

Through Heidi, I met a young sun-worn Aussie woman who's working for a mining company (not the same one that's giving Heidi money). She had just spent three weeks in Western Australia looking for Gouldians and says she saw in excess of two hundred. But she had no luck whatsoever in finding their nests. Still, the gold-mining company will be happy with her numbers; and who knows what figure will appear in their reports?

Jean and Tim are joking with me that they don't like the name I've given to this field site, Cootamundra. They insist that it be called Cowabunga. I tell them that I'll give them each a copy of John Williamson's tape, the source of the name.

It occurs to me that Williamson, the Australian folk singer, has a name as empty in meaning to Jean and Tim as Cowabunga is to me. This little incident is a poignant reminder that these students and I come from different generations, have different values, think very differently about Australia and its traditions. There are small parts of Australian life that I live as if I were born here. Jean and Tim see and often speak of Oz as if standing in a corn or soybean field back in Illinois.

Alan and I are sitting on the dirt in shade, our backs against the front tire of his Toyota Landrover. I say that Nancy and I have been out to Mud Springs, and we wondered if it's an Aboriginal sacred site. He says he doesn't know; he can't be sure about the identity of any sacred sites on the station. He's inquired of Aboriginal leaders in the area, and those connected with the Aboriginal Northern Land Council in Darwin. He's asked them several times, he says, and each time they've told him they aren't ready to say which sites had special meaning in the Aboriginal Dreamtime. They'll tell him after he puts in a new fence or bore, Alan says. His sarcasm roils my black tea.

He explains: Why should the Aborigines make noise about trespassing on sacred ancestral ground when by doing so later—after costs have been incurred and improvements are in place—they can claim the improvements and get better concessions, make political hay out of insensitive whites trashing what's important to natives?

I ask what kind of success he has with the four Aboriginal stockmen he employs.

"This mob is good. Others can be pretty bad. Depends on how strong their leader is."

Kim, the most senior of the lot, sets a high standard and the others try to be as effective. Without Kim, Alan would have chaos; before long they'd quit or he'd have to get rid of all of them.

Kim and those who follow him and work for Alan are from Halls Creek in Western Australia. They first worked at Newry in 1989. They would've worked the next year too, but Kim was caught rustling cattle from some fellow Aborigines and spent part of the year in jail.

Again and again I tell the students the names of trees and plants, how they have been used by Aborigines, when they're looking at species that are indicators of land abuse—and they do not hear me. Or they give me an unmistakably disdainful look. I know that I'm not mistaken, because a day or two later after talking with one of them, they quiz Terry, or Heidi, or one of the rangers about the very trees or plants about which I informed them. They ask in my presence, shamelessly. Jean is the most blatant. When she asks Terry—and he is the one she invariably asks—I wonder if she is not only trying to make me less credible in his eyes but also get him more interested in her than he already is. And why not? Of late, there's obvious competition between Nat and Jean for Terry's attention.

Nat caught a green frog that came up out of the hotel sink before dinner a couple of nights ago. She named him Freddie. She walks around the hotel kissing him, talking to him, cuddling him, bringing him to everyone's attention. This evening she put a yellow finch band on one of Freddie's toes. Maybe tomorrow she'll put one on her toe.

I tell Nancy and the students to return no later than noon. I want to get to Kununurra early. I have a lot to buy, the laundry to do, and I've asked Nancy to come along. It's been a while since we've had some time together away from the students. I want to stop at the Hotel Kununurra and have a few beers, hold her hand, make eyes at her.

It's now well after one o'clock, and I find myself wondering out loud if Nancy has forgotten about the shopping trip, the hour or two we'd hoped to spend alone over beers.

Amy, once again having trouble getting Cole to eat, says, "I don't know why, but I'm worried. They shouldn't be this late."

"Maybe they had an unusually successful morning at the traps," I say.

I start feeding Cole, every other spoonful a scoop of ice cream, the latest trick in my bag of spur-of-the-moment wonders to get him to eat.

They finally arrive, nearly two hours late. Nancy explains that she was driving back from Chloe and when they came to the difficult creek-bed crossing, they got stuck in sand. Neils was at the Chloe trap site, and Nancy walked back to ask him for help. He refused. He said he was busy watching a Gouldian nest and didn't want to miss a filming opportunity. Nancy reminded Neils that it was only two days ago that I'd changed the tire on his Landrover not far from the same spot, this while he stood around and watched because he didn't know what to do.

"Neils broke the first rule of the Outback," I say. "You never leave anyone stranded."

Jean says, "Neils and Terry have probably been told by Heidi not to spend any more time than necessary helping us. That's why he didn't want to help Nancy and Nat."

"That's not the story Nancy just told," I say. "I also doubt that what you've said is true. In any event, it's irrelevant. Under no circumstances do you refuse to help someone in the Outback. Heidi would be the first to shove that rule down your throat."

Tim says nothing. His face is blank. He eyes the doorway, looking for an escape route.

Nat mumbles something under her breath that I can't understand. Then she says, "Maybe Neils had some other reason for not helping us."

"Did you hear what Nancy and I just said?" I say. I feel a fire in my gut. I can't decide whom I'm angriest at, our crew or Neils.

The next morning I stop at Three Rivers to drop off the three students so they can set the walk-in traps. In the last couple of days we've gotten as many as twenty blackhearts in one of them. Since we're not doing any mist-netting on this side of Gouldian Mountain, these data are critical to understanding the range of individual birds. Before the students get out of the van, I say, "From now until the end of our stay here, no one is to invite Neils to the hotel without talking to me first. No exceptions."

The students look down at their feet or stare out the side windows. There are no questions, no reactions, nothing. Nancy breaks the long silence by reminding the students that we'll pick them up at six-thirty.

Sunday's a welcome day off, and it arrives bright and warm. At breakfast Jean and Nat outline a plan to hike west several hours. But then Nat says she's tired. "Let's wait a week or so," she says.

The conversation turns to the best way to kill the thousands of ants that live beneath the broken cement on the veranda. They've bitten all of us at one time or another when we're barefoot. Everyone thinks the ants will be with us for the length of our stay, or until someone comes up with a surefire method of extermination.

Tim has tried to drown them by flooding the area with water, but that didn't work. Now he has a new idea. "Borax should do it," he says.

"No, that won't work," Nat says. "My mother's idea is the best. Use hair spray." She looks around and sees nothing but shaking heads. "I didn't bring any, either," she yells. "What are we going to do *now?*"

"I had to grit my teeth not to say something," Nancy says.

She'd opened the freezer door, and ice cream and meat and ice trays came flying out and landed on the floor. Jean walked by as the ice hit Nancy's feet and scattered like someone had dropped a bag full of marbles. She just kept walking.

Nancy adds, "Earlier in the day it was the same Jean-for-Jean and forget everyone else. I asked her to get chairs for the group for some observations. She got one for herself, the best one. I had to remind her that there were others besides herself."

After dinner Jean disappears and Tim and Amy go for a walk west, beyond the cattle yards and the landing strip. Nat helps with the dishes and the cleanup. As I'm sweeping the floor, she blurts, "We've been talking about going to Kakadu National Park."

"When?" I ask.

"Well, whenever we can. We'd like to go soon."

Nancy, standing nearby, turns to me and her mouth drops. I take her arm and say, "Let's get out of here before I explode."

Nat overhears me and leaves without saying a word. I turn off the hotel lights and grab a flashlight. Nancy picks up Cole and we head for our tent, and privacy.

What's going on? I wonder as we thread our way along the dark narrow path. Whose idea is this anyway? We made it perfectly clear to the students before coming that we'd have a busy schedule and that travel around Australia would have to be done after our field season was finished.

After we make Cole comfortable and get into our swag, I manage to gather my senses and remember all that has happened since Tim and Jean arrived. I say, "I think we ought to give the students the next week off, let them go to Kakadu."

"I'm listening," Nancy says, the words barely out of my mouth.

I explain that it means that we'll be giving the students a week's vacation for a total of two and a half months' work in Jean's case, three and a half in Tim and Nat's—if they stay the duration. This isn't part of our agreement with any of them, and it'll be costly, but it seems that some such measure is necessary to break the tension and create a more enjoyable work and living environment. Little is working that we've done to try to please the students or make for the congenial environment we planned so hard for in Illinois.

Nancy agrees with my assessment of the problem. The next day after the students return from the field for lunch, I tell them that they're free to go to Kakadu the following week. "Nancy and I can probably arrange to trap finches with Heidi near Edith Falls and in Timber Creek while you're gone," I say.

They're elated, and within an hour the mood is unlike anything we've seen since the night Amy arrived. Jean is suddenly open and warm to both me and Nancy. It's as if we've owed this trip to the students but have been too dumb to come to our senses until Nat let us know that it was time for us to get smart.

After doing a modicum of work on finch data, the students spend the rest of the afternoon and evening going through travel books. They talk endlessly about two-wheel-drive versus four-wheel-drive rentals, about whether they want to spend a day in Darwin, about taking a river trip, about whether to take tents and hammocks, and about how much money they want from me as a loan against money yet to be earned.

The next day I drive the students into Kununurra. They spend the afternoon talking to travel agents, making phone calls for their forthcoming trip. On the return from Kununurra, Jean says that Terry and not Neils should have been put in charge of Heidi's Gouldian project. Neils, I'm to understand, is bumbling, not enough of a biologist, too interested in filming the Gouldians, not at all likable. Jean says, "Terry's a very nice guy. We saw him today and he offered to let us use his car for the trip to Kakadu."

"He's really great, he's super," Amy says. "We all like him."

Jean says, "Did you let Neils know how stupid and inconsiderate he was for not helping Nancy and Nat with that flat tire?"

"I've talked to him. He understands his mistake. But until I say otherwise, he's still not welcome at our camp."

"Well, that's good," Jean says.

"Why do they brag so much about what they've stolen recently?" Nancy asks, shrugging her shoulders. She tells me of her dismay over how Amy openly brags about what she pilfered from planes on her trip to Australia: a woven cotton blanket, cutlery, and dishes. The blue blanket has become Amy's comforter. She takes it everywhere with her. She puts it on her lap when she naps. "It's my most important possession," she says now and again.

Amy, queen among thieves, gives unsolicited instructions to the others on how to imitate her feats. She instructs with the authority of an older sorority sister who's letting pledges in on one of life's not-so-little how-to secrets.

Nancy, in one of her notebooks, writes:

Heidi is coming tomorrow, big trap days ahead. Must consolidate priorities.
1. Color band. Fives and sixes [age categories] only.
2. Bleed only those sixes that Tim has time to phenotype. Phenotype recaptures from previous years before doing anything else.
3. Double-check recaptures for data completion. Make list of behaviorally important birds. Rephenotype these.
4. Additional phenotype data to score: (a) young birds bill and bib size; (b) pull feathers for growth.
5. What should we Munsell on Gouldians?
6. Munsell masked-finch bills.
7. Ask Heidi about Gouldian phenotypes, and about longtail overlap with other species.
8. Write to banding service.

It's Nat's night to fix dinner, on the regular rotation that Nancy and I established after we tired of fixing all the evening meals. As Nat has done on other nights, she takes a long shower before cooking, a "courtesy" that has not endeared her to any of us—hungry hogs that we are by late afternoon.

After we've finished eating, Heidi comes, that gorgeous warmhearted smile of hers intact, as captivating as ever. I mention the fantastic day trapping, when Nat and I got more than one hundred fifty finches. She says that last year on Newry she and two others caught almost six hundred finches in one day.

I feel as if I've just proudly shouted from a thirty-story building that I've run an eight-minute mile.

5
Business as Usual

Lately I've been fascinated by an aggressive aquatic weed, cumbungi (*Typha domingensis*), that covers much of Lake Kununurra. It forms dense, extensive mats, and its growth is not restricted to still water. It spreads rapidly by means of branching, robust rhizomes, increasing as much as 50 percent a year, I'm told.

If not controlled, cumbungi forms floating mats over large areas of water. It impedes the flow of water downriver and reduces fish yields. Cumbungi blocks the inlet valves to irrigation pumps. Because sunlight is obstructed, the pH level and the oxygen content of the water drop, and an ideal breeding ground for mosquitoes is created. In this part of Australia, "mozzies" (as they're known here) can carry the Ross River fever virus, which is debilitating to humans.

As early as 1977, Lake Kununurra was drained for two weeks near the end of the Wet in an attempt to control the proliferation of weeds. One result was the formation of swamps and the development of thick stands of aquatic and shoreline vegetation (Gowland 1981). Though fish in the lake were abundant, thousands of the smaller ones were trapped in the aquatic vegetation as the swamps drained. The catfish population increased and became dominant. Some floating weed species apparently also increased, because of the decline in predatory insects and snails and the destruction of competitive plants (*Kimberley Echo,* April 30, 1990).

By the mid-1980s, some still talked of controlling Lake Kununurra's weed problem by periodically draining the lake. It was believed that the best time would be at the end of the Dry, or just before the first rains came in November. Then, it was argued, the high temperatures would cause the most damage to weeds. But the plan was not feasible because of irrigation demands; farmers simply could not do without water for three weeks in November. And so the spread of cumbungi continued. By 1990 locals claimed that the weed was so widespread that it was almost impossible to launch a boat, that the tick-bearing mosquitoes were unbearable, and that once beautiful Lake Kununurra should be renamed Lake Sludge-unurra (*Kimberley Echo,* July 23, 1990).

Important as tourism is and will be to the region, nothing will be done to jeopardize farmers. From their perspective, if the tourist industry is worried about the effects of cumbungi on the local economy, then the tourist commission ought to invest in a weed harvester or find other means of control. Here, as in other aspects of their business, the Ord River farmers are privileged—and highly subsidized. The yearly cost to the Department of Water for chemicals to control cumbungi in the irrigation channels is about $100,000. No money is spent to control the weed in Lake Kununurra.

Attention has turned to controlling the mosquitoes that are abundant in areas covered with cumbungi. Those who have tourist trailer parks near Lake Kununurra are especially avid. Some have used malathion. More recently, a few trailer park operators have turned to spraying with pyrethrin every four days when the mosquito population is large.

If draining the lake to make it more attractive for recreational purposes is expensive, some have suggested using explosives to get rid of the weeds, at least sufficiently to create channels from shore (*Kimberley Echo,* Nov. 7, 1988). Or, better yet, why not harvest the weeds for stock feed? That idea attracted few followers because its cost would be too great compared to buying paddock hay (*Kimberley Echo,* April 30, 1990).

Much underplayed—and unimagined by most of the local population—are cumbungi's positive advantages. It protects the riverbanks from erosion, and there are fewer algae blooms than there were when the lake lost much of its water or was drained. Further, the weed constricts the river, which is advantageous to those speeding up and down in their boats; it keeps them off the rocks. Some allege that because of the cumbungi, bird life in the area has become more abundant. This is significant because nearly 30 percent of the total number of bird species in Western Australia are unique to the Kimberley region, and some are endangered (Harris 1990). The lake, along with Lake Argyle and parts of the Ord River, are wetlands of international importance.

With the students off in Kakadu National Park, we use the opportunity to trap in Timber Creek with Heidi and two young travelers from the United States.

Timber Creek is full of tourists, and they're all over us, curious about what we're doing, how we take the finches out of the nets, what they look like up close, what kinds of measurements we're taking. Despite the many questions and constant intrusions, we're very successful: more than a hundred longtails and twenty-two Gouldians on our first morning of trapping. Heidi is delighted with the catch, as is Nancy, who bands and measures wing length and beak color as fast as her hands and mind can move.

As soon as we pull up the nets, we have a quick lunch and then begin looking for patterns in the data. They suggest roughly "intermediate" beak colors, which means that the more data we can get on blackhearts between Newry and Larrimah the easier it will be for Nancy to make reasonably solid inferences about a geographic beak-color cline. And who knows what else? She discusses all this with Heidi, and before she's through we've decided that as soon as we finish here we'll go east, to Willaroo Station, for trapping. For the first time Heidi has started taking Munsell data on the Gouldians: not just on beak color, but also on head, breast, and belly feathers.

Heidi is puzzled by how few masked finches we've captured here. She says that in 1986 she banded some five thousand finches in the Top End of the Northern Territory. Of these, about one thousand were masked finches. What has changed? she wonders. More predation by feral cats? Different fire practices? She really doesn't know. Like so much of the ecology of outback Australia, the list of questions always dwarfs the list that itemizes secure findings.

We get up at 4:30 and drive the rolling, winding 170 kilometers to Willaroo. It's a cool and windy day, and we have the road to ourselves. Heidi says she's seen lots of finches at Willaroo. We shouldn't have much trouble catching what we're after, she remarks.

We set up the nets, anxious to trap blackhearts and perhaps a few Gouldians, and take measurements on their beaks and anything else we have time for. But all that come to the nets are sixty or so harshly chattering apostlebirds. Finches are nowhere in sight.

Looking for something to do, hoping I might happen on a promising trap site, I walk a long stretch of the nearby creek. I see no finches and few birds of any kind. The creek edges are steep, full of brush and trees, almost impossible for setting up mistnets even if the finches were here.

I return to find Heidi and Nancy on their knees, repairing the nets, which were attacked by a large white cat that lives on the station. After more than an hour of intense sewing by both of them, the job is more or less finished. Heidi, edgy with frustration, shouts that she'll kill any cat she sees.

It's time for lunch, and so far we've only netted doves and apostlebirds. We've yet to see our first finch. We eat and have tea with the station manager. He claims there's been a considerable increase in feral cats on the station. He shoots them whenever he sees them. He also shoots pigs, dingoes, crows.

"You're not supposed to shoot crows," Heidi interjects. "It's illegal."

He ignores her, goes on with more tales of how and where he shoots all the creatures he named. He claims that the crows eat young chickens on the station, and that unlike scavenging wedge-tailed eagles they serve no useful purpose.

Heidi makes no further attempts to remind him that what he's doing is illegal. She knows, he knows, we know that she won't press the case or even tell her bosses at the Conservation Commission, because probably nothing would happen if she did. And if she were to file a complaint and the station manager was reprimanded or fined, Heidi would undoubtedly discover that she now had an uncooperative manager. Maybe many of them. The Outback, I continually rediscover, is the largest small place on earth.

Though covered with mosquito bites and the signs of being exposed to too much sun, the students are upbeat, chatty, totally positive about their trip to Kakadu. They talk in grand generalities about all there is to see, how great it would be to go on a hundred-kilometer backpacking trip in the park. For Amy, the highlight of the trip was having their own rented four-wheel-drive Nissan, bumping and careening over boulder-strewn roads unknown at home. She said she thought often about Cole.

The first time she takes him up into her arms, she declares, "He looks thinner than when we left." I shake my head. I think he's put on weight.

Since their return, Amy's been chiding Nat for her unwillingness to sleep with Tim and her in their tent in Kakadu. Nat had refused, saying she could only sleep in the nude and was embarrassed to be seen that way. They had worked out a compromise. Jean gave up her one-person tunnel tent to Nat and slept with Tim and Amy. Nat still found herself exposed one night when it started to rain and she felt compelled to hurry out with nothing on to retrieve what they'd left lying on a picnic table. Amy says, "I don't know what Nat was worried about. Tim and

I wouldn't have done anything uncivilized in front of her. Besides, she walks around here all the time in the middle of the night nude and all the sleeping horses see her."

The low point of the trip had occurred in a Thai restaurant on their return. A waiter accidentally spilled a plate of food on Nat and apparently didn't apologize. This upset Jean greatly. The affront was compounded when the manager only reduced the bill 25 percent. Jean thought the bill for all of them should have been canceled.

The tales of the Kakadu adventure prove to be a prelude to serious matters of the moment. Nat found a dead cat at the front door of the hotel when they returned. This is a bad omen, she says morosely. Jean also came upon a bad omen; she saw a couple of red-backed spiders inside a hose beneath our sink. What am I going to do about the spiders? she asks.

I think: The same thing I'm going to do about the black pythons that love to snoop inside my blue jeans while I'm showering—nothing.

Beverley, the cook, was visibly upset when I saw her, and her descriptions would have made a humane-society sort go berserk. She'd been out with Wes and the other ringers, watching them dehorn and castrate cattle. She says she saw blood everywhere, and then something altogether different when Wes shot a wild steer in the face and didn't finish it off for four hours because he said he had something else to do that was more important.

Jean is beside me in the van on a return shopping trip from Kununurra, uncharacteristically absorbed in a travel book on Australia. For the first time since her arrival, she asks me several questions about the country. And then: "What do you know about the Great Barrier Reef? It sounds like a great place to visit."

Imagining that soon there'll be a demand for another vacation at our expense, I look to an empty space between two distant trees and say, "Look at the size of that 'roo!"

Nat's just announced that she "never feels well."

"Never?" I say, taken aback.

She claims that she has the Epstein-Barr virus and has to watch her liver. We should have known she wasn't well, she proclaims; we would have, if only we'd paid attention to how much she's been suffering from the heat (not at all, as far as I can tell), and how, of late, she's been eating huge salads with lots of cheese and fresh vegetables.

What am I to make of this claim? Is it anything more than another attention-getting fiction? Nat, in fact, has had no problem consuming ample quantities of alcohol between stomach-filling Nutella snacks. She's also gaining weight, and she has as much energy as anyone in the field.

"I'll take you to the hospital in Kununurra if you need some medical attention," I say. "Any time, just tell me when."

"I don't need *that!*"

Yesterday morning before the sun came up, a new Nat phobia arose. She confessed that she's suddenly deathly afraid of lighting the camping stove. I asked why. "I'm just like my grandmother," she said.

I lit the stove and boiled some water and made my tasty cowboy coffee. When we were out the door and ready to leave for the field, Nat said to me, "Why didn't you boil some water for my tea?"

I said, "That's your job."

She pouted the whole day.

I make a special trip into Kununurra to print out notes I've written up on my laptop. The only place in town with a printer wants to charge me sixty cents (U.S.) a page. I change my mind. When I return to the hotel I notice that the ants have broken through the new cement that Tim used on the front porch to seal them in their interior homes. It's taken a mere two days.

Nancy is once again absorbed in data. Overwhelmed is the only way to describe the mood I read in her face and hear in her voice. But it keeps her mind off the problems we continue to have with the students. Problems that occur daily. They have only been back a couple of days and already I'm asking myself, Was the Kakadu trip anything more than an expensive ice cream cone for the students, a children's treat quickly forgotten?

Cole has just gotten his third tooth, all of them since we arrived. He's definitely putting on weight, and he looks great. Despite all the dirt that he rolls around in, he hasn't been this consistently healthy since before we first put him in a day-care center. Nancy has started changing him to adult food. She buys a cooked chicken at every opportunity so she can grind it up and mix it with fresh squash and whatever other vegetables she can find among the meager offerings in Kununurra.

Nothing seems to make Cole visibly happier than afternoon baths in his tiny pink bowl, in the center of the hotel floor. He splashes water everywhere and doesn't want to get out. Amy seems to enjoy these interludes as much as he does.

A call came in from one of Nancy's future colleagues, someone I don't know. He has looked over the house that's being offered to us by virtue of Nancy's placing seventeenth in a university lottery of thirty-three new faculty. We have forty-eight hours to decide whether or not to buy the place, sight unseen. The price: $242,000. It's two-story mock-Spanish, with lots of glass and high-beam ceilings; it has a decent-sized backyard and just over two thousand square feet; we'll be on the "view" side of the street, and we'll only have to walk about ten minutes to get to our university offices. We'll have to pay twelve thousand in rent to hold the house for six months, until we arrive in Irvine in December 1991. He said it's worth buying—"a good deal, given your options." So I guess we'll go ahead, and jump from a $7,000 mortgage to one just under $200,000. All very California, and why not? What's a little paper debt?

Another burner on our gas stove has gone out and I feel the need for caffeine, so I go down to the station kitchen and find Beverley, the bottle-redhead from Birmingham with the outback tan and the impossible accent. She's just run out of coffee. She offers me tea and as many of her freshly baked blueberry muffins as I care to eat. They are delicious.

Beverley and two of her friends—one working here as a maid, the other as a painter—are scheduled to leave on Friday. She asks if I could give the three of them a ride into Kununurra on my Wednesday shopping run.

Beverley says she's enjoyed her stay at Newry but is glad she has come to the end of her contract. "I'm tired of being a zookeeper for a bunch of animals," she says. She explains that two months of isolation and no transportation and long days of cooking and cleaning up after eight station hands have left her exhausted. She needs stimulation and civilization. She's ready for the beaches of Queensland, the bustle of Sydney.

Yesterday she went horseback riding. It was a mixed blessing. Although it was Sunday, it was just another workday for the ringers—which meant that since Beverley was on horseback and in the area where they were mustering, she'd have to work like everyone else. It didn't matter that she hadn't the slightest idea what to do when a calf or steer broke from the herd and wouldn't turn and run in the right direction with a little prodding. Like everyone else, she was expected to chase down wayward cattle. These were Wes's rules, and the head stockman is a "pretty demanding bastard," she exclaimed.

No one, Beverley claims, is exempt from the reach of Wes's iron hand. Sandy, the young first-year jackaroo fresh out of high school in New South Wales, has—so far—shown amazing resilience to abuse from Wes. One morning he took his time getting up, and he discovered that that was a memorable mistake. Wes knocked him out of bed with the burning spray of a fire hose. Another morning Sandy showed up late for breakfast and learned another lesson about strict conformance to the group norm—as defined by Wes. He publicly berated Sandy for his tardiness, then told Sandy that anything that he hadn't eaten by the time Wes had finished his eggs would be thrown in the garbage can.

But Wes, Beverley said, is invariably understanding toward the four Aboriginal ringers that form part of his crew. He lets them get food before he serves himself, and he's in no hurry to reprimand them or remind them of their mistakes.

Newry Station, from the stories I've heard about other cattle operations in the Outback, is not exactly typical. Many stations employ no more than a token Aborigine, and whites find numerous ways to remind them of their second-class "blackfella" status. Even at Newry one might judge that Aboriginal stockmen are disenfranchised outsiders. At meals or when watching television, they all sit together at one end of the table. They rarely have much to say and almost never join in a conversation outside their own circle. They usually leave the table as soon as they've eaten.

As I savored one of Beverley's blueberry muffins, she recalled station rules, a litany of irritations she said she could do without. Since the day she began, she wasn't supplied with cooking oil or fat and had to get all of it from the lean range beef that's the main dish for every meal. She got staples once a week and had to make them last, because if she ran out of potatoes or sugar or tomato sauce before the new stores came, it was up to her to invent a way to convince a tired mob of ringers and jackaroos and jillaroos who worked sunup to sundown seven days a week that they would just have to do without.

Now there was the problem of the moment. How, Beverley wondered out loud, was she going to make an edible curry out of the tough hunks of beef that filled the basin behind us? Soak them in salt...? she laughed. Got any other quick, surefire solutions? Sighing, she said for the fourth or fifth time, "I'm *not* a cook."

Before I left, Beverley asked if I'd heard about the recent incident at nearby

Avergne Station. It seemed that the previous weekend all the station hands and the cook had driven into Kununurra on Saturday night shortly after the station manager and his wife had left for their own night on the town. In Kununurra the grog-thirsty ringers bought nine cases of beer and lots of rum and coke. They returned to the station with the aim of busting loose. But in the frenzy of their misbegotten craving, they'd given little thought to the fact that the owner's twenty-four-year-old daughter had stayed behind at the station. They had forgotten, or didn't know, that she'd be all too eager to enforce the ban on anything more than very moderate social drinking—which, as at Newry, meant two beers a night and three on Saturday, no exceptions.

As Beverley told the story, the manager's daughter somehow got her hands on the liquor supply shortly after the crew returned to the station. She hustled everything over to the homestead house, locked the doors, loaded a shotgun, and waited. The ringers knew what was going on. They also knew that their tenure on the cattle station was over. The only thing left to do was to return to the bunkhouse, pack their possessions, and leave a mailing address for their final checks.

Several of the ringers at Newry had been lucky by comparison. A week earlier, while Alan and his wife, Roz, were away in Katherine, four station hands had taken a truck with the idea of going into Kununurra to "get hard on the piss." But twenty-five kilometers from town, a rear wheel flew off. They zigzagged down the road for several hundred meters before landing in a ditch, luckily avoiding a major mishap, and spent most of the night waiting for a ride. It was several days before the truck was repaired and they got the good news and the bad news. The good news was that Alan hadn't caught them drunk and acting like fools, so their jobs were momentarily secure. The bad news was that the axle repair bill would cost them some $A1,500.

Driving into Kununurra two and three days a week, I find it impossible to ignore the Aboriginal presence and their treatment at the hands of whites. The more I see, the more curious I become, and it leads me to mining local sources to find out what I can about their history in Kununurra.

Aborigines came to Kununurra shortly after the town was established in the early 1960s. Their numbers didn't increase significantly until 1968, when they began receiving wages for work on cattle stations comparable to those received by Europeans. Previously cattle stations had given Aboriginal stockmen and their families little more than rations of tea and tobacco and the occasional bag of groceries for their efforts. With the dramatic rise in costs associated with the Pastoral Award System of 1968, which gave Aboriginal station hands wages comparable to whites, and at a time when the cattle industry was discovering labor-saving devices (particularly helicopters), stations no longer wanted to employ Aborigines or allow them to live on their leased lands (Gibbs 1984).

Once Aborigines were effectively exiled from a system they had helped to create and maintain, their way of life began to change dramatically. No longer were their lives subject to a predictable seasonal rhythm, alternating between a six-month Wet and an equally long Dry. The Dry had been occupied with mustering cattle, mending fences and bores, branding, sorting, and shipping livestock to market. The Wet, by contrast, had been a long unpaid holiday known to Aborigines as "Big Sundays," a time for initiation rites, ceremonial enactments and reenactments, and the settling of disputes (Shaw 1980, p. 266).

Most Aborigines who came to Kununurra moved into a settlement two kilometers northeast of the town center. Known as the Kununurra Reserve until 1986, it had only six houses when the Mirima people arrived from nearby cattle stations (Bolger 1988). Seven houses, a community center, and an ablution block were added to the reserve in the 1970s. These, however, were token additions, and the facilities were inadequate. A typical house had three rooms, no furniture, and a wood stove. There was no indoor water. Old and young alike had to walk a hundred meters to get to a toilet. In some houses there were as many as twenty-five people. Most Aborigines used the houses for storage. They lived outside in tents or slept under trees. During the Wet the reserve population would more than double, thereby putting an even greater strain on the settlement. In addition to the crowding, problems arose because different tribes were mixed together, which led to antagonism and fighting (*West Australian*, Aug. 7, 1973).

Other Aborigines had to do without anything resembling real housing. They lived in squalor beneath tin or stick lean-tos, in burned-out car bodies, and in makeshift tents along the edges of Lake Kununurra and Lily Creek. Most were infested with lice; many suffered from eye diseases. They drank water from filthy pools. They couldn't or wouldn't join others in the reserve, not so much because of a shortage of housing but because of tribal and family differences (*West Australian*, Oct. 23, 1986).

Some Aborigines who moved into Kununurra were placed in houses in the town proper by the Department of Community Welfare. They were not consulted on how they felt about living among Europeans. This was an indication to some Aborigines that the welfare agency had no interest in spending money on adding to and upgrading the reserve. The perception of neglect abetted neglect (Taylor and Burrell 1981).

Housing for Aborigines outside the reserve was as inadequate as it was inside. A 1976 town census showed that there were seven Aborigines per dwelling, compared with three in houses occupied by Europeans. The State Housing Commission was in part responsible, since it was the sole builder of houses for Aborigines. As of 1982, the commission had constructed only forty-one homes for them, clearly an inadequate number. Aborigines accounted for more than 25 percent of the town's population. It was a youthful population; more than half were under the age of ten (Taylor and Burrell 1981; Dames and Moore 1982).

In the 1970s the Western Australian government pursued a "salt-and-pepper" policy. The aim was to mix Aborigines with the European population, not just in Kununurra but throughout the state. In Kununurra a problem arose because of the desire of the government to allocate three or four houses for Europeans to each one built for Aborigines. This required the housing commission to build many more homes for Europeans than it had thus far, or would do in the years ahead. The issue was exacerbated by resistance among Europeans in Kununurra to the policy and to occupying houses adjacent to Aborigines (Taylor and Burrell 1981).

Europeans complained that Aboriginal households were disproportionately large, and they noted that the problem was compounded by the frequent arrival of kin who lived nearby, often on the same block. In European eyes, Aborigines drank too much and they had far too many loud and disruptive kin gatherings and parties. They showed little or no concern for the appearance of their homes and their yards.

Pressure from Europeans forced local housing officials to segregate Aborigines. The task was eased by the government-sponsored purchase of cattle stations and the development of Aboriginal outstations, both of which were attractive to some of the town's Aborigines. Still, many Europeans were not satisfied. They thought that too much Aboriginal "riffraff" had stayed in town, notably those who fought and drank to excess in public.

In one of the two open-to-the-public town bars, Aborigines were blatantly excluded from an outside patio. The rationale was that they were unruly and left unsightly messes. The owners seized on the opportunity to turn the patio garden into an extension of a "European-only" indoor bar. The method of exclusion was the same as that used throughout outback Australia: posted dress and behavior codes, selectively enforced.

Over the years Europeans have expressed a great deal of resentment over the free social services made available to Aborigines, goods and services the Europeans are not getting. They're indignant about Aborigines getting free housing, free water, free electricity, allowances to the parents of Aboriginal children attending school. No such monies are available to European parents, and to many this is evidence that Europeans, not Aborigines, are the real victims of discrimination in northern Australia.

Europeans in Kununurra frequently denounce Aborigines as "dole bludgers," people content to live off government welfare. More than half of Kununurra's Aborigines are unemployed. They are aware that the wages in most of the occupations open to them—and these are few indeed—are not much higher than unemployment payments. Their social norms and individual needs aside, this fact acts as a disincentive to seek employment. It is a situation that is not likely to change significantly in the near future, and one that is implicated in a widespread belief that Aborigines will continue to play an insignificant role in economic development in northern Australia.

Amy and Nat talk frequently of wanting a real Midwestern thunderstorm: the heavy rattle in the sky, the drenching downpour, the instant puddles to wade through. We're now deep into the Dry and have had nary a drop of rain since they arrived. Home beckons frequently for the students.

Both women have been nagging me about not having hot-water showers. After the second complaint, I said that if hot showers were that important, they could gather wood and heat up the rusted iron "donkey" outside the hotel. But this, clearly, is too much trouble. They think it my job to do it for them, better yet provide them with all the comforts of home.

The other day when they mentioned the cold showers, I reminded them that they invariably take their showers in the afternoon when the water is warm, that I alone do so early in the morning before the sun comes up, when the water feels near freezing. They scoffed. My comfort is not their concern.

The weekend upon us, Nancy and I decide to take Cole and drive into Kununurra to get a motel room and splurge on a Chinese meal and a decent bottle of white wine. And then enjoy each other at the midnight hour when Cole is fast asleep.

As we're leaving, Jean asks if we'll deliver a message to Neils and Terry at Keep

River. I say we'll be glad to make the twenty-minute detour. She hands me an envelope.

In the van I say, "The envelope's sealed. Looks fishy to me."

"Relax," Nancy says. "You're getting too suspicious."

"Mutant Polish genes," I laugh, tugging at a familiar explanation for behavior that strikes her as beyond the pale.

We have a great time at dinner, and afterward. Returning Sunday at midday, we find Amy and Tim lying in hammocks, drugged and roasted red from the harsh sun. Nat and Jean left early and went for a hike toward the low-slung mountains to the west. They told Tim they'd return by sundown.

While the four of us and Cole are eating dinner, Neils drives up to the hotel.

"Any idea what he's here for?" I say to Nancy.

"No idea. I didn't invite him."

Neils gets to the front door and peeks in. "Sorry I couldn't make it for the party last night. I had to do some more filming and I got back too late to get here."

Nancy looks over at me and I look over at her. What's this all about? I'd made it perfectly clear to everyone that Neils was not to be invited to our camp unless I gave the okay. Now we're hearing that as soon as we turned our backs, the students not only invited him to come around but had us deliver the invitation!

I ignore Neils, don't ask him to sit down or offer a beer or food as I do to all our guests. I keep eating, my eyes on my food. Neils says he wants to talk to us again about the behavior of blackhearts and Gouldians; he needs more information for his feature-length film in progress.

"Six weeks from some Sunday," I say sarcastically and keep eating. He gets the message.

Before Neils is through the station gate in his 4 × 4, Nancy says to Amy and Tim, "What's up? Rich told you at Three Rivers that Neils was not welcome here. Rich was very angry and didn't leave any doubt about how he felt and why. You not only went against what he said, but you did it behind our backs."

Amy says, "I thought it was okay to have him over if you were gone." Her face suddenly shows more red than her scorching from the day's sun.

Tim bows his head and picks up a spoon and moves it toward his empty plate. He doesn't say a word.

"Who invited him?" I say.

"We all did," Amy says. Tim nods. Then she says, "We really invited Terry, but Neils could come along if he wanted. We didn't see anything wrong with that."

I get up and walk out, not sure what to do.

When Jean and Nat arrive an hour later, I don't allow them to sit down before I tell them Neils has been by to explain why he didn't come to the party last night, and we want an explanation of why they invited him.

"I really didn't know you didn't want them here," Jean says.

I stare her in disbelief for a long moment, and then say, "Are you deaf, or do you have Alzheimer's disease?"

"I *really* don't know why," she says.

I shift to lecturing mode. "You, if all the stories are half true, have done as much remote back-country backpacking as any of us. If you learned anything from that experience, you know how important it is to be able to depend on others. Now you're trying to feed Nancy and me poppycock to the effect that you don't know why I was so angry with Neils, and why I made him persona non grata around here. Furthermore, you, as I recall, pointedly asked me before going to Kakudu whether I'd told Neils how I felt about what he did, and you sounded every bit as censorious as I was. Now you're trying to make Nancy and me believe that you haven't a clue about what's going on. What do you take us for, two-digit morons?"

Nat says, "I thought he couldn't come here anymore because he came one night uninvited."

I turn to Nancy and move in close, and I whisper in her ear, "When were they born?" Turning to Nat, I say, "Why don't you go find Freddie and give him a big kiss?"

Nat says, "I really don't think it's a big deal what Neils did to Nancy and me."

I pick up Cole and say, "Nancy, let's get out of here before I lose it."

At Cootamundra I'm walking down the winding trail through thick sorghum grass with Nat when she sees something on the ground and turns to me and earnestly says, "I want to be a frog." Her big brown eyes, magnified by her thick glasses, meet mine. "I *really* want to be a frog!"

I now think I understand why she fears snakes.

Jean's eager to pass around some pictures she's just gotten in the mail. They're of a bronze bust of Eddie. "Isn't it great?" she asks Tim as she hands him the photos. Tim, who never met Eddie, smiles and nods, then passes them to Amy.

"Oh! Is that what he looked like? I never knew *what* he looked like. I sure wondered!"

Jean says, "Eddie would have approved, he really would have. He always said he didn't like the idea of leaving this life with nothing."

Amy hands the photos to Nancy, who glances briefly at them. Nancy shows no emotion, says nothing, passes them to me.

I turn them into strong light, stare at one, then another. Eddie in bronze looks like a Greek Olympian: handsome, proud, with long golden curly hair, much older than I remember him. My bleached pictures of him are different: standing on the roof of his garage beside me with my chainsaw, cutting a huge branch; his growing beer belly heaving as he tells an off-color joke; responding with intelligence to a question about bird behavior that Nancy has posed . . . I continue staring at the photos, thinking that the bust bears only the vaguest resemblance to the person I'd known. I look over at Nancy and read in her face that she feels as I do, that there's something terribly tasteless about the bust—its rendition as much as the idea behind it.

I return the photos to Jean. She lovingly moves them from one hand to the other. "I don't remember the crow's-feet and those lines on the left side of his mouth," she says to no one in particular. "No, I'm sure he didn't have those. They could have made his hair a little shorter too." She picks up her left leg and brings it into the hammock and begins swinging left and then right, obviously lost in reveries.

The photos of Eddie open Jean up a bit. The day after receiving them, she ruminates out loud about him at length. Although she knew that he had a bad heart, she claims she rarely thought about the future or what might happen to him. The future just wasn't an issue. Eddie's family has lots of money, and they would take care of everything.

Getting tired of the long eulogy for Eddie, Amy brings out a familiar paperback entitled *Life's One Hundred or Two Hundred Most Important Questions,* something like that. Questions such as: Would you eat a handful of live cockroaches for $40,000? Would you be willing to die exactly fourteen years and two days from now if all those years were utterly blissful and whenever you felt lethargic you could take a pill that would immediately make you feel marvelous?

On one of our twice-weekly shopping trips into Kununurra, the students spent most of the hour's ride reading questions from The Book. One brief exchange addressed the following question: What would you do if you knew you were going to die in twenty years?

"I would live life to the fullest and go wild every day with my Visa," Amy beamed.

"If you did that, you'd be leaving all kinds of debts for other people to pay after you died," Nat said.

"That would be *their* problem!" Amy said.

Nat then told two stories about how she'd not bothered to return money when she had been given too much change by a salesperson.

Amy had a similar story, but for more money. Our resident Bonnie, minus her pistol, teamed with an acquiescing Clyde.

I'm busy cooking and Cole's crying and needs something, maybe a little attention. Everyone but Jean is outside, or taking a shower, doing I don't know what.

"Jean, would you mind?" I say.

"Mind what?" she asks.

"Could you give me a hand with Cole? He needs something, maybe just a kiss on the cheek."

"I want to finish this chapter first," she says. And reads on.

Nat makes a stir-fry on her turn to cook dinner, to which she adds an impossibly hot Indian curry sauce that she's brought from home. No one is able to eat the curry. Instead, they turn to leftovers or something in a private pantry to get sated.

Seeing that Nat is crushed, Tim comes to her rescue. "That's it!" he says. "Nat's curry is the solution to our ant problem."

He takes a gob of the deep orange sauce, goes outside and stuffs it into the large cracks the ants use, then says, "Watch!" The ants continue to move about as usual. Additional applications of the unpalatable curry have no noticeable effect.

In the morning, the curry's gone, consumed by the voracious ants, we guess.

At Dingo Creek, sitting on a boulder waiting for finches to come in, I turn and look back at the cardboard table. Nancy's intently measuring the color of a long-tail beak. Tim's taking a leg measurement with calipers. Nat's holding up one of the rectangular wooden boxes full of birds and trying to decide which one to take out next and kiss before giving it to Tim. Jean's expertly drawing blood. A familiar scene.

I stare at Jean and wonder if what I see is the image I'll have of her in five or ten years. Not this person I've come to know less and less by the day, as I wonder incessantly about the motives for her behavior. What am I to make of what anyone might see sitting where I sit? The purple and blue running shoes, the oversized yellow aviator's glasses worn in sun and in shade, the gold wedding band on the middle finger of the left hand, the firm thighs that show no signs of aging, an attractive full face with fine teeth, and the wide-brimmed Victorian sun hat, and—most of all—that innocence promised.

Alas, more than once, I have been seduced by the appearance of innocence, what it promises and does not come close to delivering.

Beverley and Susan and Susan's boyfriend, Peter, ready to leave Newry for good in the morning, come to the hotel and ask if they can buy some beer from us. There's now nothing to lose in offending Alan and Roz, they reason. Hoping to turn them away, I say that we have nothing that's cold and that in any event all anyone has is light beer. Everyone but Nat understands the meaning of my words. She shouts that she's got a case of Castlemaine XXXX, then follows with, "How much are you willing to pay for what I have?"

"Whatever you want," Peter says.

For the next several minutes Nat talks out loud to herself and to the other students and to Freddie the Frog, wherever he might be. She can't remember exactly what she's paid for the beer and this is causing her considerable distress. She starts looking to everyone to help her so she won't lose money. Finally, overcome with indecision and her own alcoholic needs, and much to Peter's chagrin, she says, "You can have six cans, no more."

Later, long after Nat has turned gleeful at her small profit, Nancy and I resolve that we'll remind the students again that no one is to accept anything of substance from the cooks, and that no one is to sell any alcohol to the cooks or station hands, or buy it for them on our runs into Kununurra. If the kitchen help and others want to accompany us on a trip to town and buy alcohol for themselves, that's okay. Small courtesies for others here are not worth the risk of endangering our stay at Newry.

Needing stimulation, I've been looking for ways to engage the students in conversations that have some intellectual bite. Little seems to work. The other morning before turning to my daily journal, I read a provocative essay in the *Weekend Australian* by the iconoclastic journalist Phillip Adams, on the reception given to *The Silence of the Lambs* by critics around the world. Adams lambasted critics for describing the movie as art. He thought it was a "meretricious piece of sleaze,... toxic waste," a kind of insidious pornography that numbs our senses.

About a week ago, Amy and Nat and Jean had said they'd seen the movie. So I made a special effort to bring the essay to their attention. Amy had no interest, but

Nat and Jean read it. Nat couldn't get beyond the fact that Adams was wrong about how the sadistic serial killer, Buffalo Bill, loosened up a victim's epidermis before removing it. Adams, Nat ranted, didn't understand the biology of skin loosening and therefore anything else he had to say had to be wrong. Jean got stalled at the identical juncture, which prompted her and Nat to reinforce each other. They concluded that Adams could not possibly have seen the movie.

"But what about his reactions to it?" I asked Jean.

"I thought it was really good entertainment," Jean said. "It was pretty funny also."

Nat said, "I thought there were some real funny parts in it too. They really made me laugh."

"I liked it a lot," Amy said. "I love to be scared, and it was real scary."

We're trapping at Tanya this morning. The nets are busy, the bags full, we're shorthanded. Nat's in one of her work-as-slow-as-possible, bird-talkathon moods: "Just a minute, sweetie, and you'll be free. You *poor* little thing! I hope you're not losing too much blood. Are you feeling better now? Slurp, slurp, slurp, drink!" I'm in one of my intolerant moods. I ask Nat politely a couple of times to quit spending so much time adoring the birds and talking, to stop turning them upside down and kissing them. The third time I say something to her, she says, "I don't feel well."

"Why didn't you say something earlier, then?"

"I've had to pee all morning."

Nancy's eyes and mine meet, and then we both stare at Nat. She moves her head nervously from side to side. After several long moments, she gets up and goes to relieve herself in the bush.

When she returns, I note again how unpredictable the finches are in coming to a particular water hole. To enliven the quiet morning, I try to get the team members to speculate on the causes of this unpredictability. There are no takers until Nancy begins weaving a story, speaking slowly, leaving plenty of room for others to join in. But no one does. Nancy starts:

> I think the birds are moving from water hole to water hole for two reasons. Their first and most pressing need is to outwit the predatory birds that derive their living from other birds' need to quench their thirst. Insectivores and other meat eaters may not have to drink regularly. They can get lots of water from their diet. But most seed eaters can't subsist entirely on metabolic water, especially when they're breeding. The seed eaters also need to assess surface water availability in order to know when it's time to give up and leave for wetter places.

In Alice Springs we baited traps with seed to attract zebra finches. We found we couldn't make the finches really trap-happy, that all birds foraged elsewhere regularly. Moreover, even the birds that ate frequently from our seed baits didn't appear to start to breed until seed was generally available, weeks after the drought broke. I think that the unpredictability of conditions over much of this continent has selected for continual assessment of ecological conditions, even when they appear favorable. I bet these birds sometimes travel miles to get their morning's water, even when they know that water's close at hand. It'd be very interesting to know how the small seed eaters decide which water hole to go to on any given day, and which species, if any, serves as pointman.

Historically, much of the surface water available in this area at the end of the Wet was in the depressions in the riverbeds, the billabongs. As the dry season progresses, the water level in billabongs drops, and fewer and fewer water holes are available. The distances between alternative water holes increases, and so I'd guess does the pressure of the predatory birds. Eventually the distance between watering holes gets so great that the finches leave to go north, where conditions are wetter. I bet they're following the riverbeds. Someone could radiotrack them on horseback to see where they go. Perhaps the routes they follow are quite stereotyped. Maybe the young learn a migration pattern from the adults, like other birds and vertebrates. Maybe birds follow this pattern so rigidly that the bill-color divergences we see are the result.

This idea that birds are sampling water holes over large distances, even while they're breeding, may explain why we catch so many birds. A water hole that persists for any period of time is a magnet. Birds will return again and again to see how it's doing, to see if it has dried up. The information they gain goes into some complex calculation that tells them when they had better leave.

I joke about Nancy's newfound interest in optimal foraging. Then I comment on the parallel with bird feeders in the northeast of North America, where the regular food provisioning by humans has resulted in some formerly migrant species becoming year-round residents.

"That could be happening here," Nancy responds. "But if so, it's in a very early stage. We're catching some of the same birds at distances around a kilometer. According to Heidi, the birds are leaving at the end of the Dry, despite bores and cattle troughs and lawn waterers."

I say, "Do you really think that migration along river routes could explain beak-color radiation in longtails?"

"I know what you're thinking. If that's the explanation, why shouldn't we see similar patterns in other bird species? The longtails' closest relative, the masked finch, doesn't show this trend. We've found remarkably little variation in masked finches, wherever we've trapped them. Nor do other birds show this trend. Obviously, some birds are distributed more locally. They seem to specialize on using the "soaks" that might be more permanent sources of water. But if they're more sedentary than longtails, they should show even greater phenotypic divergence. Then, like the budgies and most of the small doves, some birds appear to be much stronger fliers than finches. They may be able to make it to reliable water holes at much greater distances and therefore may not need to rely on traveling the riverbeds. I admit, the picture's pretty hazy, and there are plenty of holes in this argument. Yet I do think it would be worthwhile to follow the longtails up the riverbeds when they leave the breeding grounds. I'm sure that's where they will go."

This afternoon Alan invited Nancy and me to his home for a Saturday night barbecue. It's a weekly event for the station hands and, as we soon discover, a special dress-up evening for Newry's resident Aborigines. While others come in thongs and shorts and singlets, the Aborigines wear pressed slacks, colorful snap-button cowboy shirts, and shiny pointy-toed cowboy boots. They sit together in a circle in darkness on the grass, silent and alone, still social centuries away from all the others who slouch in chairs and on benches at tables on the patio.

The food is what one might expect on a station where the manager is not the owner and has a family and is expected yearly to increase profits. Everywhere are signs of frugality and expenses forgone. There are a couple of ordinary salads, bread bought in town, beer, and beefsteaks that come from a range cow slaughtered at a bore by two of the ringers. I have two small pieces of steak to get my protein fix, and to reaffirm whether cooked rare or well done they're like all range beef that's never known a feedlot. The answer's a clearcut yes. Nancy can't finish her steak; it's that tough.

Alan invited us because he said he wanted to see one of our bird books. He has only the vaguest idea what longtailed and Gouldian finches look like, he declared. We were delighted to accept his invitation, even though we were uneasy because the students weren't included. We told them that we felt an obligation to go, that they shouldn't feel slighted. But we also saw the barbecue as an opportunity to get to know Alan and Roz better and to ask some research-related questions.

The most urgent question on our minds involves confirmation of the rainfall figure that Heidi has given us for Newry. We also want to examine rainfall records for past years. This last week Nancy worked hard at the painstaking job of analyzing and graphing big chunks of the computer printout data on blackhearts that Heidi had given her. The analyzed data have provided Nancy with a great deal more precision—and some major corrections—relative to the vague verbal generalities that Heidi has repeatedly articulated since our arrival. The monthly trapping information by age class and year has raised numerous questions about breeding behavior in Nancy's mind. Some of them will be clarified if she can get more precise data on rainfall in the vicinity of where the birds were trapped.

Alan said that, since the beginning of the Wet in October 1990, his single rain gauge at the station house has registered fifty centimeters. The condition of the grasses around Gouldian Mountain indicates that the rainfall there was somewhat greater, but how much greater he can't quantify. Nor can anyone else. Rain gauges are few and far between in this part of Australia.

From what we're able to learn from Alan, however, rainfall outside Newry Station, especially to the west and south, has been considerable—perhaps three times greater than at the station. From a pastoralist's point of view, rainfall figures don't mean much. What matters is the timing. If rain comes at the end of the Wet (in February and March), then there'll be more feed for cattle.

Some of this information comes as a surprise. Heidi told us that Newry's rainfall had been no different than the rest of the Top End. She said that this would be a very good breeding season for longtails and Gouldians because of the unusually heavy rains. And heavy rains, in general, they were. Newspaper and TV reports went wild with claims that the Top End had had its wettest year in recorded history, more than two hundred centimeters. Now, however, we were being told that not only was the rainfall at Newry less than a quarter of this figure, but that according to Alan it had been a below-average year.

We can't do anything about the fact that there's only one rain gauge on the station, or that it's more than twenty kilometers from the longtail breeding grounds. The best Nancy can do is make some adjustments based on the trapping records and on Alan's memory of years in which the rainfall at the study site has been different than at the homestead.

Our reconnaissance work led us to believe that there's plenty of sorghum on and around Gouldian Mountain. Now Alan tells us that cattle eat so little of the two varieties of sorghum grass that constitute the major portion of finch diets that it would be impossible to consider birds and cattle competitors. Cattle eat some sorghum when it's quite young, and again when it resprouts after a fire. Otherwise they leave it alone. Thus to a practical-minded scientist, an assessment of food availability for the finches should be a low priority. And that is how we've thus far treated the issue.

"Has Jean gotten back to you about doing a research project on foliage nests? Or anything else?"

"Not a word."

Nat's got all of her luggage firmly secured with thick rope—to keep out snakes and other things that crawl, and perhaps the rest of us who she fears might snoop when she's away. I've caught her checking the knots, wanting to be sure that nothing's been tampered with. As if to remind herself of the precautions she's taken, she enjoys ribbing Tim for the luggage that got lost, his "stupidity" is not putting name tags all over everything.

When Nat's not around, the other students joke mercilessly about her peculiarities and her neuroses. They particularly enjoy making fun of her fear of snakes, the tape around the inside bottom of the all-net tent that she keeps insisting I replace with a stronger variety, and the misshapen steel cot she's put her sleeping bag on to avoid the snake she just knows will some night come after her. I've reassured her several times now that her tent is the most secured fortress against snakes anywhere in the Outback. She's not convinced. She seems to need the warm blanket of yet another fear, one made palpably real by constantly talking about it.

The longer we're here with the students—and it's now close to five weeks—the more Nancy and I feel like outsiders in our own home—one we designed and built, and for which we pay almost all the bills. Notwithstanding our mighty efforts before coming, and our attempts to communicate on their terms since the students arrived here, everything about their behavior tells us that they've decided it's us versus them. Jean is the most blatant in this regard. I've lost track of how many times she's offered one of the students a beer or some of her potato chips, whatever she might be eating, never once extending the offer to Nancy and me, who are sitting nearby or perhaps beside her. Maybe the slight wouldn't be as noticeable if we didn't make such an effort to be inclusive: not charging them for food their first several days with us, my often buying them beers when we're in Kununurra, doing their laundry, taking care of the accounts, giving them the van every other weekend to do as they wish, offering all of them beer or wine or whatever Nancy or I bring from the refrigerator.

Leaving the bank, I see the four students sitting on a parapet outside the post office, lined up like thirsty cows at a water trough. They're completely absorbed in their mail. Rarely do I witness such singular intensity in this group. At such times I think that only prisoners who infrequently receive word from the outside world are as needy of reassurances that they haven't been forgotten.

Letters, cards, letters, and more cards. Ten, twelve, fifteen are given me by the students to be mailed whenever I make a shopping trip into Kununurra. Two weeks into her stay, Amy had written some twenty-five letters. From the beginning, she has described the trip to the post office as "the best time of the week." When I see the stack of letters that she receives—six or seven some days, and the gleam in her eyes as she reads them, and reads them again, and then reads them out loud for everyone's appreciation, I think I know why. Whenever Cole's asleep or Amy has a free moment, she's writing to someone. I have no idea whether she has a hundred friends, or ten to whom she writes repeatedly, two and three times a week.

On an average Tuesday or Friday jaunt Nancy and I mail or receive only one letter or card. Some weeks there's nothing, or at most a single piece of university correspondence that Nancy feels compelled to answer.

Sometimes when I go to the post office and pick up the students' mail, I go through a list of reasons why we're so different in our writing habits. Nancy and I are too busy trying to keep the research running smoothly to find time to write. We no longer have friends or family who can't wait until we return home to hear about our latest field adventures. We're both simply negligent, and I in particular am finally—after years of writing fifty to seventy-five letters a year—slipping into a pattern characteristic of all but a very few friends, a habit that in effect says don't write unless in need or there's great and exciting news that absolutely has to be conveyed.

At times I conclude that there's no one that either of us feels we have to impress by letting them know where we are or what we're doing. Other days, dissatisfied with all the reasons that have crossed my mind, I conclude that we'd be writing as much as the students if this were our first trip to Australia. But then I have to correct myself. We didn't write much at all when we were in central Australia for our first extended stay in 1986–87.

There's a somewhat similar difference with respect to taking photos and filming what we're doing. Though Nat brought along Nancy's new camcorder, we haven't used it at all. I've taken some 35-mm black-and-white shots and a few slides, but nothing like the amount taken by the students. They all have 35-mm cameras and a bag of extra lenses and lots of film. The quantities of film take up much-needed space in our tiny hotel refrigerator.

Nat and Tim brought enough film to keep them shooting at sunup and sundown and throughout the day for months. On her arrival Nat announced that she expected to take an average of seventeen shots a day. And she has. So, to a somewhat lesser extent, have Tim and Jean. Shots of flowering trees, green frogs coming out of our kitchen drain, white spiders crawling over food cans in the hotel, the students lounging in their hammocks, cattle being herded into the yards across the road, the van beside the hotel or parked alongside the road, anything and everything that will allow them to put together a slide show for friends and family on their return. Just the kind of evening entertainment that Nancy and I have seen too many times and now dread having forced on us when we're invited to someone's house for dinner.

We're sitting around the campfire and the sky has turned magenta. I cannot imagine a more tranquil evening. Tim and Amy are remembering friends from high school days. Nat's poking the fire, trying to move a rock here, another one there. Jean's looking off into space, into the darkness south of the cattle yards. How I'd love to hear Jean sing and play her guitar. She has a charming voice. But she has yet to play it since she bought it.

Cole, playing on the ground between Nancy and Jean, dirty from head to feet, picks up a stick and hands it to Jean. He makes a cute giggling sound, wants her attention. Everyone chuckles and smiles. Jean, expressionless, gets up and without saying a word heads toward her tent. How profound and unfathomable the mysteries of the mind, endlessly twisted by history.

I take Jean and Tim and Nancy to Tanya. They're going to do some nest observations, I'm going into Kununurra to buy supplies and food for everyone. We wait several minutes at the site for Terry and Neils to arrive, to pass on some information. As we're waiting, I notice that we have a flat tire on the right front. I bring the problem to everyone's attention. Nancy joins me, saying, "What can I do to help?" Tim and Jean turn their backs and walk away.

Not eager to show my ire at their unwillingness to help, I tell Nancy not to bother. I get in, slam the door, and drive back over the boulder-strewn road to get away from everyone, caring not at all that I may destroy the tire.

Parked in the gravel that edges the Victoria Highway, as I fix the tire my Buddhist brother comes to mind. I think of how much I envy his patience, his apparent tolerance. But for how long would he be patient and tolerant here?

Birds are abundant in the Ord Valley, which isn't surprising, I suppose. The water of the lower Ord and the feed that grows in the vicinity invite breeding, large numbers and diversity. This tropical habitat was made even more inviting by damming the river and opening the area to irrigated farming, thereby creating still more water surface and abundant food attractive to birds. In the semiarid tropics, a dry season checks the buildup of a population; this doesn't occur in an area that's more or less continually irrigated. In the Ord, population explosions, if not the norm, are not unexpected.

Most abundant of the new arrivals to the Ord since the river was dammed have been water birds, now estimated at around two hundred fifty species. Most of them are migratory—perhaps as many as 70 percent—and are covered by international treaties (Storr 1980). Among the birds that most affect farmers are little corellas (*Cacatua pastinator*), magpie geese (*Anseranas semipalmata*), and brolgas (*Grus rubicundus*). Little corellas, long targeted as the valley's major avian pest, are notorious for feeding on sorghum. They attack the seed heads, and the plants are often broken when the birds alight on them. They also feed on rice, sunflower seeds, rock melons, watermelons, and sweet corn—all abundant in the valley (Beeton 1977). Magpie geese are attracted to new horticultural plots, those recently plowed, which makes it easy for them to extract roots, tubers, and rhizomes. They're also drawn to grasses that come up after the first rains of the Wet. Brolgas, which eat chickpeas, maize, and sorghum, cause their greatest damage by trampling grains.

When regions neighboring the Ord experience poor rainfall, birds from afar seek sustenance in the relatively stable water- and food-rich environment of the lower Ord. In a dry year, as many as a thousand red-tailed black cockatoos (*Calyphorphynchus banksi*) will migrate into the Ord Valley. In one unusual year (1979) an estimated eighty-seven hundred redtails were seen, allegedly the largest single gathering of these birds ever observed in one location in Australia (*Kununurra News*, June 23, 1979).

Some conservationists are now concerned that the redtail population is threatened by angry farmers, who shoot or poison them. Their numbers are endangered elsewhere in the state because of aggressive colonization of their nesting sites by little corellas.

Because reliable data are difficult to obtain, and seasonal variations make easy generalizations suspect, no one is sure exactly what the balance sheet on Ord avifauna looks like. Some claim that despite habitat destruction and poisoning and shooting by farmers and sporting activity, the birds are winning—their populations are increasing. Others contend that while a few species have probably benefited from human activity, most populations have suffered (Beeton 1977).

Whatever the verdict on changing numbers of birds in the Ord Valley, they've been seen as a significant hazard to farming since the late 1940s, when crop research was initiated. Records at the Kimberley Research Station show that between 1947 and 1951, 70 percent of experimental grain crops were written off because of pest damage. Ninety percent of this damage was attributed to birds. More than twenty years later, it was claimed that at the onset of grain production, flocks totaling up to thirty thousand birds were consuming two tons of sorghum a day. In this particular case (the year was 1970), however, the extensive damage was not just due to the local population. Below-average rainfall in inland areas of the Kimberleys caused a massive seasonal in-migration into the Ord. A similar event may have occurred in 1973. In that year, by one estimate there were fifty-six thousand little corellas in the Ord Valley eating some five tons of sorghum every day of the year. In addition, an estimated thirty thousand magpie geese were even more destructive than the little corellas; they trampled the sorghum crop. The geese are difficult to chase away and, unlike little corellas, they're not classified as "vermin" and therefore can't be shot (*West Australian,* Feb. 27, 1973).

In 1982 an Ord farmer with forty hectares devoted to rice said that he lost 20 percent of his crop to an invasion of thousands of magpie geese. He claimed to have spent three to four hours a day chasing them away. The geese were quick to adapt to new threats, taking only two to three days to adjust to blasting radios. Even though it's possible to grow two rice crops a year in the valley as opposed to a single crop farther south, the goose problem—and a glut of too much rice—will force growers to turn to other crops.

Troubling as these large bird populations are to farmers, the case against birds is easily overstated. A 1981 estimate of grain intake by birds in the Ord Valley was 322 tons a year. But of this amount, only 19 tons was consumed prior to harvest; the rest was waste. Thus the birds were not nearly as serious a threat as the gross figure consumed suggests. Still, birds do break off grain heads and they facilitate the spread of disease, losses that are hard to quantify but which may be considerable (Andrew et al. 1985).

Because of the size of the bird populations and the apparent cost in crop losses, well into the 1980s it was more or less open season on little corellas. The attitude was to remove the birds as quickly as possible, and by whatever means: explosives, poison, baiting, just about anything that came to mind as a possible control method. Finally in 1986 the Western Australia Department of Agriculture began issuing "damage licenses," permits that allow farmers to shoot or dispose of birds or other animals that are causing losses. Under this provision, the farmers are rarely allowed to kill more than a dozen birds or half a dozen wallabies at a time.

Damage licenses or not, it's always been difficult and costly to control the corellas and other birds. Looking for new solutions, many farmers have turned to largely nonlethal methods: scaring them off with sirens and shotgun patrols. In many cases, however, all that farmers are doing with their shotgun patrols and sporadic killings is releasing frustration, momentarily satisfying themselves that they're coping with a ever-present problem. Corellas learn quickly, even to the point of recognizing the cars or pickup trucks that bring in the shooters. Among the more effective methods of keeping a large flock away from a crop is to identify the "scout birds" and shoot them. But this approach is usually only effective for a couple of days.

By 1990 a few farmers were using helicopters to "herd" little corellas and move them to a distant site. Some farmers with tropical fruit and nut crops use a net mesh to protect their crops. A method once employed, and now out of favor, was to create a diversionary feeding area and attempt to train the birds to use it.

Ord Valley farmers are generally given the benefit of the doubt by government conservationists about how much damage they're inflicting on native animal populations. Apart from the politics of power, the bias favoring farmers is abetted by the argument that little is known about the population dynamics of these so-called pest populations. Working in favor of the birds is the size of the populations—probably larger than they would be without the irrigated farming. Also aiding them is the fear that a significant slaughter of any species, even those classified as pests, would create a public outcry that could change matters in a manner highly unfavorable to any control measures used by farmers.

Because of an almost universal ban on the export of Australian fauna, the government is missing out on a major source of income to be gained from the huge corella populations. A controlled export market could bring in millions of dollars for the Kimberleys. Exporting "undesirable" birds is attractive not because of losses suffered by farmers, or problems that arise from a native wildlife export policy, but because the Australian government favors killing over export.

I leave the hotel in search of Ali, the young English woman who's recently been hired to clean Alan and Roz's home and the kitchen where the station hands eat. I'd promised to lend her an anthology of the best writings—prose and poetry—on the Territory (Headon 1991).

The few times I've talked to Ali I've been impressed by her intelligence. She's nineteen and plans to enroll at an English university in the northern fall, to study literature and Italian. When I first met her I'd judged her to be a university graduate with some worldly experience. She seems much more mature than any of our students. The few times she's come to the hotel, she's been uncomfortable. Later she's told me that she found the students' conversations tedious, Nat's attempts at risque jokes offensive. Nor does she feel comfortable with the ringers, whose crudeness, in her mind, begins with their faulty English and extends to their tolerance for dirt and their ravenous appetite for protein. She tries to bring a modicum of civilization to their lives by putting fresh flower arrangements on the long cafeteria-style table where they eat.

Despite Ali's sometimes uppity English attitudes about the Australian colonials and uncouth young Americans in her midst, she has an amusingly perverse sense of humor. She proudly wears a gold heart-shaped locket. Inside is a picture of her deceased grandmother's lover. Ali loves to tell stories about this infamous scandal that's part of her family tree.

Nancy too is impressed with Ali, not so much from talking to her but more from noting the superb job she did last Saturday when she joined us at Dingo Creek out of curiosity. "Much better and faster at recording data than Nat," Nancy told me later. Ali was also good-humored and didn't complain. "She's just the kind of person we need." That afternoon we talked about trying to recruit her, pay her to do some of the work that the students are doing that isn't up to what we want. I approached Ali the following day. She said she wasn't interested. She hates the flies and the heat and derived no pleasure from working with the birds. The onetime morning adventure was more than enough to satisfy her curiosity about what we're doing.

When I find Ali, she's busy with Roz and Alan's children, Simeone and Marcus, helping them with their daily lessons. Roz is in Darwin for a couple of days to attend a yearly educational conference. The conversation quickly turns serious. Ali has just finished *From Here to Eternity* and wants to know what I think of it. I confess that it's been so long since I read it that I hardly remember enough to discuss it intelligently.

Ali goes into the kitchen to make some tea, and when she returns she looks positively ashen. I ask her what's wrong. She's just seen a couple of cow tongues lying on top of several slabs of beef in a cardboard box. "Now I know I'm going to be a vegetarian," she says resolutely.

I leave her and start back to the hotel when Magda, the new cook, stops me and invites me to enjoy some scones that she's just made. We sit at the kitchen table and she tells what she thinks about station life at Newry. Like the previous cook, she complains bitterly about the narrow range of foods she can prepare, the lack of eggs and fresh vegetables and other basic necessities. On Curtin Springs Station, where she worked some years ago, the ringers would go crazy if they couldn't have bacon and eggs for breakfast.

"What kind of meals are you cooking down there?" she asks.

"Little bit of everything," I say. "The students like chicken." I describe Jean's great chicken curry.

Her eyes light up, the deep lines on her round cheeks tighten. "The blokes here would go crazy if they heard there was chicken or fish around. They'd die to get something besides beef."

She says she's getting by with a toaster that only takes two pieces of bread at a time. She has to use a broom handle to keep the oven door closed when in use.

The conversation turns to "whinging," an all-encompassing Australian term for complaining. She says that, like most station people, the ringers bear their fate and deny the need for help. They stoically suffer the consequences of whatever befalls them. "The whole left side of Wes's face was swollen up when he came in this morning. He says he was playing with a pimple or something on his face and it got infected. He won't say a thing more about it or go to a doctor. It'll either go away or he'll just accept what happens. Going to a doctor is like using sunblock. All these blokes see it as pampering. I could tell them that Australians have the highest skin cancer rate in the world and they wouldn't listen. Never."

On the way into Keep River, Neils pulls up alongside me and says he wants to talk. He's agitated and noticeably angry. He and Terry had an argument in the laundromat in Kununurra and Terry hit him in the face. He doesn't explain the reason for the fight, says only that the rift was long coming. He really wants to leave right away and return to Denmark. All that keeps him here now is his filming project on the Gouldian finches, he says.

Ali came to the hotel today with a magnificent gift for all of us. She'd found a piece of discarded plywood measuring about fifty by a hundred centimeters, painted it white, and then, drawing on her memory of Old English script, lettered in black "Blackheart Hotel." I was speechless, as was Nancy. After we recovered, we thanked Ali for the treasure she'd created for us. Jean, Tim, and Nat were there, but they showed no interest in Ali's work and barely acknowledged her presence. There was no curiosity about where she'd be going when she left Newry Station.

Nancy and I got Ali to pose with her sign, and later we hung it on the south end of the hotel for all to see as they came along the station road to the main buildings. Ali's singular gesture, and the growing symbolism of the sign, was not lost on either of us.

Since we arrived at Newry, we've done all of our grocery shopping at Charlie Carter's in the small town's only mall. From the beginning, I've been struck by how few Aborigines shop there. They all go to the Tuckerbox, about three blocks away and on the periphery of the town center. I've gone to the Tuckerbox once or twice, but haven't returned because the offerings are so meager. And because the store's cluttered and claustrophobic.

The local Aborigines shop at the Tuckerbox because the white store owners will give them credit and cash their checks, even greet them on the street. White customers are allegedly reproved by the store owners for complaining that Aboriginal customers smell or look as if they've been rummaging around in a mountain of garbage—which they often do. All this contrasts sharply with the treatment of Aboriginal peoples at Charlie Carter's, where their trade is discouraged and where their checks are not cashed. When I asked why they weren't, a manager claimed that bad checks from Aborigines sometime in the distant past killed credit for all of them. I asked if he ever got bad checks from whitefellas. He said he didn't care to answer any more of my questions.

It's hard to know how much those who own the Tuckerbox see Aborigines as equals and how much their motives are driven by economics. Since the arrival of Charlie Carter's in the late 1980s, the Tuckerbox has lost a lot of business. Charlie Carter's is not only larger and more upscale than the Tuckerbox, but it offers more variety of food, it houses the town's only news agency, it's adjacent to several other businesses, it's brighter and roomier and more inviting, and it rarely has more than the occasional Aboriginal customer—a big plus in a town as blatantly racist as Kununurra.

Jean has had an extensive rash on her stomach for more than a week. Neither she nor anyone else has any idea where it came from. She's worried, but she doesn't want to see a doctor in Kununurra for a diagnosis, as I suggested. She's also got a reddish sore on her face, below her left eye. That one she's had for almost two weeks. She claims she got it from wearing sunglasses. Based on its location, I doubt it; I think she was bitten by something. But I say little. She is easily offended when someone offers explanations alternative to her own.

Today was a successful day trapping. We stayed unusually long at Dingo Creek, until three in the afternoon. We got ninety-eight blackhearts and six Gouldians. The recapture rate remains quite low—consistently low, in fact. This suggests that the longtail population is larger than our early estimates indicated. But then, how large is the "local" longtail population? How many of them are just "passing through"?

Amy comments that she's now in the habit of calling Tim "Cole," and that Cole's on her mind all the time. She really does seem to have fallen in love with our little boy, and of late she's been talking about wanting three or four kids of her own.

Her snooping catches me off guard at times. And it doesn't always please Nancy. This afternoon Amy was listing the names of everyone from whom Jean had received mail over the last week or so. Amy's once voluminous mail has suddenly dropped off, enough to make her declare that if all those she's communicated with don't write soon they won't receive any more letters from her. Nat said almost the identical thing just two days ago. But Nat hasn't received much mail. This is well known to Amy, who has a running tally of exactly how much mail everyone has received.

Terry and Neils have supposedly mended their broken fence, though to judge by Neils's repeated references to Terry as a dickhead, I sense that Neils only pretends to be amiable in order to finish his filming. He has now come forth with the reasons for the fisticuff. It all began, he says, when Terry dropped a stone down a nest full of Gouldian nestlings and killed two of them. Later that same day, Terry filled the Landrover with gasoline; it takes diesel. Then on toward evening, Terry broke the only washing machine to which they have access at Keep River. So Neils has to take his clothes into the laundromat in Kununurra, which he finds unacceptable. Somewhere in this run of mishaps, Neils let Terry know that only dickheads are this incompetent, and Terry, an Aussie from birth and no doubt with a picture dictionary full of images of people whom he has seen called dickhead over the years, responded in the only way he considered appropriate.

It's Sunday and a day off for the team. Nancy and I have an early breakfast alone, then decide to do some trapping at Chloe, as much to be together as family as to give the students the day to themselves. What a fantastic surprise we get! We set up the familiar tight V with the mistnets and within the first half hour we trap thirty-eight Gouldians, thirty-four of which are unbanded juveniles. This kind of catch—the number of Gouldians, to say nothing of those that are unbanded juveniles—is simply unprecedented. It's a day, we chime to each other, that will rank high on our list of all-time special experiences.

Amy gave Nancy the grocery bill and told her what our portion was. Nancy looked over the receipt, shook her head, then carefully went through the groceries and checked them against the receipt. She came over to me and said, "Now they're embezzling money from us. It happened last week and I didn't say anything, thinking it must have been an honest mistake. Looks like the students have come up with yet another way to save money on their food bill."

I rechecked the amounts against the box of groceries that were for us, and then against the communal groceries. Sure enough, we'd been overcharged nearly twenty dollars. I nodded to Nancy and she took the bill to Amy, who was milling around with Tim and Jean. Nancy pointed out the discrepancy. Amy turned red and stuttered. She finally said she didn't know how it was possible that a mistake had been made. Nancy suggested that they recheck the bill against the groceries together, to see that she herself wasn't in error. Amy had no interest in doing so. Nancy said nothing about the previous week's discrepancy. Nor did I.

Cole has conjunctivitis. Both of his eyes are red and puffy and weeping. He looks positively awful. My guess is that it's from all the dust and dirt here at our camp site. Everyone feels bad for him. Is there a correlation between how he looks and how he feels?

As soon as we saw that the conjunctivitis had flared, we called off all fieldwork and took him into the hospital in Kununurra. Nurses there didn't think that it was nearly as bad as it looked, perhaps because they see such terrible cases. Their descriptions of eye infections among Aboriginal children are enough to scare any normal mother to death, Nancy no exception.

It's comforting to have the hospital at Kununurra so near at hand and so accessible. Whenever we've needed to take Cole in, the hospital has been open, and there's never been more than a half-hour wait to see a nurse or a doctor. They are friendly, well trained, and a lot more approachable than medical people we've dealt with at home. There's no attempt at intimidation through technology, titles, attire.

Even without insurance, a visit is affordable, and the price of a prescription is ridiculously low by American standards.

We had a moment alone and I asked asked Nancy what she'd tell other ecologists about taking students into the field.

"Don't pay them," she said.

Lately I've noticed that Jean regularly checks my accounts to see that I've reimbursed her to the penny for food and other expenses she's paid out of pocket. Two weeks ago I accidentally shortchanged her fifteen cents. She promptly brought it to my attention. This morning I went back over my own numbers and found that I'd overpaid her several dollars recently. Since she checked my little red-and-black book just yesterday, she must have seen the error. She said nothing to me.

In the beginning there was trust without suspicion between us and the students. Now that trust is fractured—nay, gone.

It's four days since we first noticed Cole's conjunctivitis. It's hanging on, and both Nancy and I worry that it may be more severe than the nurse led us to believe. Appearances are driving our reactions. Nancy's renowned patience, I've noticed, is not so renowned when it concerns her only child's welfare.

We take Cole back to the hospital to have the conjunctivitis checked. His eyes are still weeping. We are given reassurances that it doesn't go away as quickly as we might like, and we shouldn't worry. We won't, will we?

Alone and away from the students, we use the occasion to go to the Chinese restaurant. We have honeyed king prawns, which are delicious, and a nice chardonnay.

Yesterday was Cole's first birthday, and it was to be special—as special as Nancy and I could possibly make it. We'd bought a bakery cake in Kununurra, presents for Cole, funny hats to wear, silly horns to blow. We'd bought expensive food and elegant wine. We even decorated the hotel. By suppertime at five in the afternoon, Cole was perched in his blue car seat on top of the dining table, at the king's end. We were ready to celebrate. All that was missing were those whom we'd invited, our team. Surely they'd arrive any minute, because they'd been told in a hundred ways and a thousand times that Cole was the center of our universe, and that his first birthday was to be memorable.

Six o'clock came. Nat and Tim and Jean, and the van, were still nowhere in sight. Amy was fidgety, Nancy worse. The students had never been late for a meal, and they had rarely worked past five. Cole was hungry. He had to be fed before he fell asleep. We could wait no longer.

What happened? A flat tire? An engine breakdown? Tim couldn't be found because he didn't know the time?

Six-thirty, seven, then seven-thirty—and still no students. Cole had eaten a little, dozed off, and now he slept. I fumed and began to lose focus. My eyes turned inward. I could no longer see Nancy, or Amy, or even the near walls. I began hearing a jackhammer pound inside my head. Reason was on the wane. I would not—little beasts in my head were now saying—listen to reason unless my head could fit through the hole in the tire, or the broken axle was dropped in my lap and I fell to the floor from the obvious weight. All I wanted now was more of the wine we'd bought for the party, all of it.

Now in the haze of my day-after anger that still rages, will rage forever, I see the three of them coming through the open door at eight o'clock, laughing, joking, carefree, utterly indifferent. I see Cole, his head aslant, his eyes closed, snoring a baby's snore. I hear Jean or Nat or Tim say to Nancy they didn't think it was "any big deal" what time they got back. Oh, there's a party, that's right, I forgot, I hear one of them saying. Okay, but I need to shower first, Nat says. I want to eat something, one of them says.

Was there a problem with the van? I asked. A flat tire? I asked. Got stuck? I asked. I was this polite; reason hadn't totally fled after all.

No, they replied to each of my questions.

You had a problem with some birds? I asked. They kept coming in? I also know I asked.

No.

Why, then, are you so late?

Their silent stare said, We don't have to account to you.

I reached for another bottle of red wine, opened it, brought it close, offered it only to me. Feeling rage that I have seldom felt, asking—never to know—why oh why *this* too?

I think Cole was awakened, finally. And I think once or twice I saw Nancy smile. I remember this or that voice saying something or other about a gift or two for Cole and isn't it cute.

Now I remember just sitting at the long metal table after everyone had gone to their tents, my head in my hands, then my forehead kissing the metal top, wanting to cry. Maybe I did, all over inside. I don't remember how I made it back to our tent. Maybe Nancy came and got me.

In the morning—it was about 4:30 A.M., my watch said—I jumped up and grabbed my pounding head. I put on shorts and stepped outside and took a long pee. I charged up to the hotel. I grabbed the radio, found the loudest station, and turned up the volume as far as I could, hoping to wake every living thing on the station. Most of all the students whom, save Amy, I felt like taking by the neck one by one and saying, Where were you born, in a barn among snorting hogs?

Tim was the first to arrive, hands to ears, eyes half closed. Sheepishly, innocently, he said, "What's up?"

With a voice meant to override the radio, wake Alan and Roz, get horses and 'roos and wild cattle on the run, I began my hectoring, lecturing monologue. I ran on, I'm sure, with a fury of street language that left no doubt in Tim's mind that I'd been brought up in the meanest ghetto in America.

I accused. I judged. I condemned. I wanted all of them to pack up and be gone before sunup.

Nancy spent more than an hour with each of the students today. The great calm voice of reason, always seeking reconciliation. War by other means, even when war is war. We live by different moral imperatives.

She spent an hour with Tim, walking down to the station gate and beyond; more than an hour with Nat, into the parkland field where the wallabies roam; time enough and then some with Jean out beyond the cattle yards; too much time with Amy up to the Rec Room and back. Trying, in a way that only she can, to get through to the students: about what we had all agreed to before coming, about their behavior since coming—the lies, the betrayals, the lack of interest and cooperation, everything. Trying to speak to each of them on what she imagines to be "their turf." Asking as politely as possible: *What* is going on, and *why* are you behaving this way?

Nancy doesn't always tell me everything. Especially when I'm like this—too hot to touch. But she did tell me she thinks they "understand." She says that Amy and Tim and Nat seem eager to reduce the obvious tensions. But Jean, as she has from the time of her arrival, stubbornly holds out, insisting there's "no problem" worth discussing. Nancy said it's transparently clear that the others are following Jean's lead, lemmings all. The implication was clear: without Jean no problem, or a minor problem, or small problems easily resolved.

"I'd like you to help out," Nancy said to Jean.

"I don't see that as my role," Jean answered.

Obviously, I hadn't heard her that day at Montezuma's when I asked if she'd like to come and she said she didn't want to baby-sit. What I hadn't heard was that she didn't want to help with anything; she only wanted to pursue her private itch, her disruptive agenda of the moment.

Later Nancy says to me, "The problem is, none of them is mature enough to handle the situation by the guidelines we drew up. They're stuck somewhere around age fifteen, in unforgiving adolescence. On the one hand, they want us to be indulgent parents who do everything for them, but they resent our having rules, structure, any expectation of their behavior. And the ringleader is a spiteful witch to boot! She made sure that they would get home late on Cole's birthday."

Nancy bursts into tears for the umpteenth time. After she calms down, she says wearily, "Maybe we should send them home."

"No," I say. "My investment is too great; yours is too great. We won't come back. The breeding season is too far along. We can't get anyone to replace them—easily. We will lose too much. We must tough it out."

When will I say, *Enough!*?!

Amy came looking for me, and found me talking to Alan outside his home. He was telling me that "blood finches" (crimson finches) eat other finches. I made no effort to counter the wild claim, and when the conversation lost steam I turned to Amy, who had Cole in her arms. I took Cole and gave him a big hug and kiss. Alan said he had work to do and left.

Amy said, "Can you come right away? There's a snake inside."

On the fast walk back to the hotel she said she'd seen the snake in the old cook's quarters, where everyone save Nancy and me were storing personal effects.

"How big is it?" I asked.

"A meter or so long, and it's crawling among Tim's clothes. It's brown, it's black . . . I don't know what color it is."

"How long did you stay and watch it?"

"Oh, about fifteen minutes."

"You had Cole in your arms?"

"Yes, of course."

I guessed that the snake was probably a black python and harmless, like the ones I often saw in the bunkhouse bathroom. ·

I took the lead as we cut across the low brown grass and the pool of water below the water tank. A thin dark object a couple of steps in front of me caught my eye. I thought it was a piece of rope or piping, then changed my mind when the object began to move, finally disappearing into a humpy mound of dead grass.

Once inside I grabbed a broom and poked around among the suitcases, clothes, books, everywhere Amy had seen the snake. I looked in the shower, where two days ago a fist-sized tarantula had walked across Nancy's foot. My eyes fell on the large crack between the bottom of the metal wall and the ground that ran along the north side of the hotel.

With all the water around the hotel from the leaking pipes and the open kitchen drain and the shower water that overflows, chances were that we'd see several more snakes in the hotel before we finished our work. The chances were poor, I told myself, that we'd see any king browns. The chances had better be poor, I told myself, knowing that Cole is often crawling around on the floor or playing in his pen while Amy is engrossed in writing yet another letter decorated with every imaginable kind of doodle.

Terry, Nat and I are working at the data table at Chloe, banding and taking measurements. We have several bags full of birds on the ground, some there for forty-five minutes or longer. The nets fill with birds, we clear them, putting the Gouldians and blackhearts into calico bags. Terry and I return to the table, Nat follows with a willie-wagtail in her right hand. Holding it by the feet, she takes out her camera, changes lenses, then asks Terry to take some pictures of her and the bird. While Terry fiddles with the camera, Nat adores and kisses her most recent object of affection.

"If you don't mind, Nat, later for photos and affairs of the heart," I say. "We've got lots of birds to process and it's hot."

Terry says, "Let her do whatever."

"Mind your own business. She's working for us, and I happen to care what happens to the finches."

Nat goes back to work, quickening her pace. Terry fumes the entire morning.

We return from the field to an uncommon mess in the kitchen. Some papers (nothing very valuable) have been shredded, cups are lying on the floor, and there are some unusual droppings beside boxes where we store supplies and data. The mystery is on. Was it one of the great bowerbirds who had come in the front door that we never bother to close? Or had it come in through the old "dog window," the large square opening just to the right of the kitchen sink, where we keep the clean plates and dishes? Maybe it wasn't a snoopy bowerbird after all. Now and again a magpie lark comes through the front door or the window and is trapped for several hours, unable to find its way out. It'll sit above the stove or play among the kitchen supplies atop the tinny storage cabinet with the sprung door hinges. When I'm at the hotel alone during the day and this happens, I make no effort to get the magpie out and into its natural habitat. It will, I always imagine, find the light and make its way home before dark.

Several magpies consistently hang around the front of the hotel. Sometimes I don't even acknowledge their presence, taking them for granted as it's so easy to do with all the magnificent wildlife in our midst. But then in the afternoon I get anxious to see one of them give chase to three or four little corellas, thinking—most times when I see this happen—that magpies just love to tease and snip at a cockie bum, to remind them who's the real boss at Center Camp.

The cattle yards across from the hotel are packed with eleven hundred fat, bellowing cattle that have been driven in from the mountain slopes on the station's northern boundary. They're being sorted, branded, dehorned, castrated, and tested for tuberculosis. The test isn't reassuring, in that it's accurate only about 50 percent of the time. Like other stations in this part of the Territory, a mere fraction of 1 percent of all cattle on Newry are eventually shown to have TB. If other cattle are infected, it's almost impossible to detect because of the way the disease hides within the animals. The ostensible reason for testing under these circumstances is that other countries will not buy infected cattle from Australia. This stance is rather ironical because in many of these same countries, most in Southeast Asia, high rates of tuberculosis infection are common.

Still, there's probably not a cattle station owner or manager anywhere in the Territory who objects to mandatory TB testing. It's almost a certainty that all heartily welcome it. The federal government pays a station five dollars Australian for every cow tested, and ten dollars for a second test, which is routine. Since the testing is done at the time the cattle are mustered for sorting and sale, the quick and easy procedure (performed by a veterinarian paid by the government) involves no extra cost to the station. On properties like Newry—which is below average in size in the Top End—testing for tuberculosis is one fat cash cow.

As her familiarity with blackhearts grows, Nancy is becoming more impressed with the differences between populations. It's not just a matter of beak color; bib size and tail length also vary. Most intriguingly, Nancy now feels that she can reliably ascertain the sex of most adults at Newry in a quick glance. Males have longer tails and bigger bibs than females. In other populations, such as those at Larrimah, confusion reigns. Bibs are smaller, tails somewhat shorter, and there's no obvious correlation between tail length and bib size.

Nancy is intrigued by this difference. Some years ago she wrote a scientific paper about the significance of "sexual indistinguishability," namely, the phenomenon she is observing at Larrimah. In that paper she argued that, contrary to the conventional wisdom, species that are indistinguishable to humans may also be indistinguishable to themselves, and she provided experimental evidence that pigeons, *Columba livia,* cannot discern the sex of other pigeons they have never met before (Burley 1981). She suggested that individuals living in gregarious species may benefit from concealing their sex, or "lying" about it. Under certain ecological circumstances, females might benefit from looking like males: if female assistance improved the success of territory defense, and if males were somewhat better defenders, perhaps as a result of larger body size, then a female that defended the territory and looked like a male might be even more successful as a parent—assuming, of course, that she could get a reasonable male to take her as a mate!

Nancy can see no obvious ecological differences between conditions at Newry and at Larrimah that could explain the difference in sexual dimorphism that she sees. Unfortunately we can't spend more time in Larrimah to look more closely. In zebra finches, males engage in more nest defense than females. We don't know whether this will prove to be true for blackhearts, but it does seem likely. At Newry we've observed that paired birds spend a lot of time together, and both are frequently at the nest. Perhaps single-parent nest attendance is, of necessity, more common in Larrimah? If so, Nancy reasons, wouldn't one expect the population to evolve more male-like traits, larger bibs, perhaps longer tails? Instead their bibs are much smaller than those at Newry, and their tails seem somewhat shorter.

This is a puzzle that will take Nancy some time to solve, if indeed she ever does. She tries to redirect her thought process and think of other possibilities. What force would oppose sexual selection, cause dimorphism to decline? An obvious possibility is kin selection, efforts spent helping kin to survive and reproduce. Perhaps there is more nepotism in the Larrimah population, and even more kin mating, than in Newry. Mating of biological kin decreases the intensity of the "battle of the sexes" and should make mating partners more cooperative. It might be prohibitively expensive to examine this possibility in nature, but Nancy thinks she could explore it under laboratory conditions. She has already found that zebra finches have mating preferences for first cousins. She could examine the relative strength of a preference for kin mating in birds derived from the Larrimah population as opposed to the preference in the Newry population, and document kin interactions of all sorts in captive breeding populations. To do that, she would have to know the genetic as well as the social relations between birds. A big project.

Nancy is also struck by how "unornamented" longtails and their close relatives the masked and black-throated finches are in comparison to their next-closest relatives, zebra finches and double-bar finches. Double-bars are monomorphic and not colorful; but their plumage patterns are complex, full of strong lines, subtle barring, and polka dots. Male zebra finches have five markings—some colorful, others strongly patterned—that females lack. By contrast, longtails are models of elegant simplicity. Nancy wonders if this simplicity evolved through active mate choice or is the result of some unknown constraint. The only way she can think of to get a handle on this question is to repeat some experiments she has done on zebra finches: to increase their ornamentation and see if birds like the effects.

Much to her surprise, when she applied plastic bands of different color, as well as headdresses, she determined that zebra finches found quite a range of novel "jewelry" attractive. Further, females and males had different preferences, which is consistent with the striking dimorphism of this species. If longtails really like simplicity, she reasons, she will be less able to find preferences for novel traits if she puts them in her mate-choice apparatus.

Although she is loath to repeat the mate-choice experiments she conducted a decade ago, her attraction to the longtails is clearly increasing. Will longtails like color bands? Will preferences for novelty differ between the sexes? Will they differ between individuals from different populations? If she performs these experiments, she'll also be able to ask, Will birds from Newry and Larrimah freely mate with one another? At this point, she's willing to bet they won't. She can look to see if sex roles vary between populations in ways that might explain differences in sexual dimorphism.

A disturbing pattern has been growing steadily worse. At the end of each field day, Nancy or I ask what "interesting things" everyone has seen that day. Lately Jean and Tim seem to be in open competition to claim to have seen nothing, whereas Nat's observations have become increasingly implausible. Today she reported seeing a longtail and a Gouldian finch pair up to chase away a butcherbird that came near the tree where both birds were nesting. Nancy found the claim otherworldly. She's watched nests of zebra finches and other Australian estrildines for hundreds of hours and has never seen a single example of nest defense against a possible predator—and certainly not a predator as dangerous to a finch as a butcherbird, which eats adults as happily as it does nestlings. Gouldian finches, in fact, seem the most timid of the timid, regularly yielding to the smaller blackhearts. Nancy pointed this out to Nat, who held her ground. When Nancy commented that this was then the first report of cooperation of any sort between these two species, Nat said, Great! She now wants to send her observation off for publication to *Emu,* the Australian journal of ornithology. Nancy reminded Nat that these Gouldians and longtails have soft beaks, not fighting equipment for scaring butcherbirds, and that butcherbirds are afraid of no one, people included. You can throw rocks at them and they won't fly away, she said. In fact, we do so regularly, to try to dissuade them from foraging near our traps and nets. Why, she asked Nat, would butcherbirds flee from finches? Nat was silent.

Nancy has had enough, her legendary patience temporarily gone. She must question Nat's claims not only for Nat's benefit, but also to convey to the other students that some measure of reality must prevail. Up to now, she's tried to handle the students gently. Like most university teachers, she's learned from painful experience that students are deeply hurt if Nancy gives any hint that she feels they've let her down. Many students seem to think the mentor's role is to be uncritical and unqualifiedly supportive, a position vigorously voiced by Tim several times since his arrival.

Heidi gives me a copy of a paper that she and a staff scientist have written. It's been accepted for publication in a scientific journal. In the paper Heidi and her Conservation Commission colleague use a mathematical model to estimate the total size of the Gouldian and longtail populations at various sites on Newry. They arrive at the astonishing conclusion that the Gouldian finches number substantially more than the blackhearts. This conclusion is directly contrary to what common sense tells us, and to what all our data show. Wherever we trap, the blackheart populations are three and four times larger than the Gouldian finch populations.

Heidi and her mathematically minded colleague arrived at their startling, counterintuitive conclusion because the model determines the population size based on the recapture rate. The lower the recapture rate, the larger the population, predicts the model. The rate of recapture for Gouldian finches is significantly lower than for blackhearts, thus Heidi and her colleague conclude that Gouldian populations must be larger.

The worthiness of the conclusions derived from any mathematical model, not least one such as Heidi and her colleague have used, depends on the legitimacy of the assumptions. Wrongheaded or irrelevant assumptions give wrongheaded or irrelevant conclusions. In this case, the important assumptions are that there is no migration (the population is sedentary) and that mortality is minimal or absent. Under these conditions, given sampling with replacement (we don't remove the finches we catch, nor do we inadvertently cause their death), recapture rates should index total population size.

But no one knows anything about the migration behavior of either the Gouldians or the longtails. Nor is anything known about mortality rates. Heidi and her colleague do not make any distinction between adults and juvenile Gouldians, even though they have the requisite data. All of these factors, and others such as avoidance of trap sites and alternative water sites, could greatly influence recapture rates of the two species.

Nancy and I are astonished that Heidi, who has been trying to get the Gouldian finch included in various endangered species lists around the world, has seen fit to publish these claims. Does she believe conclusions that contradict everything she sees when trapping longtails and Gouldians?

When I caught Nancy this afternoon in the hotel kitchen, I asked what she imagined might be Heidi's motives for publishing these obviously erroneous claims. Jean, lying in the hammock outside, overheard me and charged into the kitchen to berate me for being critical of Heidi's paper. "How do you know there are more longtails?" she said. "What gives you the right to be so critical?"

"How many legs have you got?" I said.

Nancy frowned at me. She explained the huge database we've accumulated from trapping. She went back over what we thought everyone by now took for granted. She said, "Longtails are much easier to locate than Gouldians. When we find both species together, there are three to four times as many longtails on an average trap day, and roughly five times as many longtails have been banded. When we only find one species, it's invariably the longtail. Moreover, a lot of time has been spent by Heidi looking for new places that Gouldians frequent, to little avail."

Jean said, "Maybe Gouldians are living in places we don't visit or trap."

"That's possible," Nancy said. "But that's not the basis of Heidi's conclusion. Indeed, it's not a possibility even addressed in the paper." Nancy then took Jean through the implicit assumptions of the elementary population model used to predict population sizes.

When Nancy finished, Jean said, "Heidi is a field biologist and knows what she's doing. You should trust her judgment. Besides, the paper was accepted for publication, wasn't it? It wouldn't have been if what she said isn't true."

I have accumulative notes on xenophobia and its place in Australian history, in this far edge of the northern Australia where we find ourselves trapping finches. It is, I now see, relevant to an understanding of the founding of Kununurra, the damming of the Ord River, the establishment of irrigated farming to the north and south and west of this small frontier town that's been so crucial to our mission.

For the better part of a century, Australians have feared for their "problem lands" of the north, several million square kilometers of a continent that long contained only about 1 percent of the nation's population. In this northern tropical vastness, the sole developed area was a narrow, intensively cultivated strip devoted to sugarcane along the east coast of the continent. Geography as much as anything was the source of these "gnawing fears" (Younger 1970, p. 803)—the proximity of northern Australia to Asia, its remoteness from Great Britain and Europe, and an equally daunting faraway isolation from the major centers of effective settlement in the south. Beneath and reinforcing such anxieties and suspicions about northern neighbors was the White Australia policy.

Some policy it was. Formally encoded in the Federal Immigration Restriction Act of 1901, it prohibited entry into Australia of any person who could not upon command of a government officer write a passage fifty words in length in a European language. The Australian Federal Parliament proclaimed that this test was to be given only to non-Europeans (Willard 1967, pp. 121–122).

As the twentieth century began, nearly 80 percent of the country's population of 3.75 million was Australian born (Eddy 1988). It was a time when Great Britain was "home" and the Union Jack could be found everywhere, when the creation of the Commonwealth and a constitution was a "simple public act ratifying what [Australians] internally accepted, the reality of white Australia" (Crowley 1960, pp. 189–190).

From the earliest days of its enactment, the White Australia policy generated little controversy. It was simply accepted that Australians wanted no part of "contamination by 'inferior' races and sweat labor" (Crowley 1974, p. 274). It was believed that to marry and mix with those who were not white would lead to mental, moral, and physical decay (Alomes 1988). Unity in Australia meant a "united race," which implied "white, Australian and British" (Alomes 1988, p. 40). Australian nationalism and racism were inseparable.

If a sense of superiority was one rationale for what came to be seen as the Monroe Doctrine of the Commonwealth of Australia, geography, demography, and democracy were others. Australians were conscious of their remoteness from the motherland and of their peripheral position in the empire. They could also visualize the threat in large numbers: "hundreds of millions of land-hungry Asiatics," a "Yellow Peril" that threatened their democratic way of life (Crowley 1974, pp. 274, 293).

A de facto White Australia policy had been part of the Australian ethos long before it was written into the nation's first constitution. The 1890s were characterized by xenophobia and racism and ethnocentricity; the ideal of a White Australia was rampant.

In Queensland in 1878 the government threatened the Union Steamship Company with the loss of a lucrative contract if it insisted on replacing European seamen with Asians. This attitude was not unusual. The *Bulletin*, one of the country's first and most influential magazines, was supportive, arguing that it was necessary to keep Australia white to bar cheap Asian labor (Willard 1967). Australia was for Australians, and to make this point abundantly clear, its masthead read: "Australia for the White Man"(Alomes 1988, p. 32).

There should have been little reason to worry about the internal presence of nonwhite immigrants. Only fifteen hundred Chinese lived in all of Western Australia at the time of federation. Most were employed in market gardening, in laundry work, in hotels. In no real sense could the Chinese be seen as a threat, yet it was the specter of what they represented and might become that led many in this mammoth, sparsely populated state to support federation. A united conglomerate of states would institute controls on Asiatic immigration and mobilize a national system of defense to assure exclusion (Crowley 1960).

Western Australia's north was singled out. It was seen by some as the nation's front door, by others as its unprotected back door (MacIntyre 1920). More than half the size of Great Britain, the region had fewer than a thousand people at the turn of the century. A significant number of these inhabitants were Aborigines, peoples with no rights. By 1933 a mere sixteen hundred persons lived in this vast portion of Australia's north. In the next fifteen years only about a thousand were added to this number.

Throughout the 1920s it was frequently stressed in the Western Australian Legislative Assembly that the empty north should be developed to ensure the safety of Australia. The region was seen as "a serious menace to the whole Empire," the "weakest link in Australia." If this region was not settled, it was asserted, the entire country might be endangered. Only by populating the north would the country "complete the chain of defense essential for the protection of Australia against the thousand millions of colored people adjacent to Western Australian shores" (quoted in Graham-Taylor 1980, p. 38).

Preservation of the White Australia policy was one of the new nation's principal reasons for participation in World War I; it was an issue to both opponents and supporters of conscription. After the war ended, a prime minister would tell Parliament that his country had sacrificed its youth in order "to maintain those ideals which we have nailed to the very topmost of our flagpole—White Australia" (Graham-Taylor 1980, p. 240).

The militaristic maneuvering and imperialistic appetite of Japan were a disturbing specter to Australians in the 1920s and 1930s. Few believed that the Asian mainland and the islands of the south and east Pacific would satisfy Japan's expansionist designs. The only solution, many imagined, was to "populate or perish" (Markus and Riklefs 1985, p. 216). The "empty north" had become Australia's "gravest problem," because the "storm centre of the world has swung from west to east" (Conigrave 1936, p. 247).

One member of the South Australia House of Assembly was certain beyond doubt that "the 'White Australia' policy is one of the finest policies that any European community could adopt." He warned that Australians should not be blind to the consequences of the policy, especially for the country's distant northern lands. "These consequences . . . arise from the proximity to our coastline to islands and alien countries harbouring millions upon millions of coloured people. Australia is confronted on the north at no great distance by something like 900 millions of coloured people of different races. Australia must not allow extensive and attractive tracts of country to remain unpeopled along the north coast" (Basedow 1932, pp. 9, 10).

In 1933 the Commonwealth government publicly addressed the issue of the north, confessing that it had been neglected because of preoccupation with "development and settlement of the more southern portions of the Continent." Now it was imperative that the north be developed and settled "to the same standard of industrial and economic progress as has been achieved in the south" (Commonwealth of Australia 1933, p. 2). But the issue was not catching up with the more developed south; rather it was preservation of the White Australia policy (Conigrave 1936; Crowley 1974).

Attitudes and expressions of fear for Australia's northern lands were reinforced by solid popular images of Australia, images that constantly celebrated the nation's "98 percent British" population. The pride they embodied was manifested in ways large and small, as in letters to the editor that were signed "British Australian" and "British Preference." The flip side of this ethnic and racial breast thumping was the malignant belief that "the alien, the foreign and the diseased . . . might undermine simple, healthy, British-Australian harmonies" (Alomes 1988, p. 98).

The Japanese occupation of New Guinea, naval engagements in the Coral Sea, and the raid on Darwin on February 19, 1942, had a chilling effect on those Australians who had given little thought to settling the north. To those who had warned of the dangers in previous decades, the Japanese bombing confirmed what politicians had long preached: "Australia's open spaces were an invitation to invasion, and it had at last materialized with Japan's bombing" (Crowley 1974, p. 481). These worries were never more evident than in the months after the bombing when a Japanese invasion seemed imminent.

Australians began to argue more vigorously than ever for a "Populate or Perish" strategy. Prime Minister John Curtin in 1943 decided that the country needed twenty million people, then a few weeks later changed the number to thirty million. By war's end, new and more forceful arguments were being advanced for defensive development of northern Australia.

Following World War II, fear of the Yellow Peril was as widespread as it had ever been. The belief grew at both federal and state levels that every effort should be made to develop the north's potential. It was said that "all patriotic Australians would acclaim any proposal to develop our sparsely populated areas" (Drewe 1981, p. 38). If Australians did not do it, then Asians surely would.

The Rural Reconstruction Commission in 1945 strongly advocated development in the Ord Valley as a defense against the Yellow Peril. Hansards of the time were full of statements that the north had to be fortified as protection against invaders (Drewe 1981).

Economic development of the north continued to be an issue well into the 1960s. Much of the discussion revolved around misgivings about Asian intentions, especially China's expansionist ambitions. The arguments were old and familiar; unless the government took development seriously, peoples from "overcrowded" Asia would penetrate Australia's northern shores (Trengrove 1969, p. 246; Reid 1971; Aitchinson 1972; Moffitt 1972).

In the 1970s, government officials began to choose their words more carefully when addressing racial and nationalistic issues. For a decade liberal critics in Melbourne and elsewhere had been complaining that the White Australia policy was "poisoning" Australia's reputation in most of the world's capitals. In 1965 the words "White Australia" were removed from the Labor Party platform. Between March 1966 and September 1970, fifteen thousand non-Europeans were allowed to settle in Australia (Alomes 1988).

Now when the prime minister spoke of what he wanted for Australians, it was "their rightful place in their own country"; this to ensure that they were "the owners and keepers of the national estate and the nation's resources" (*The Age*, Nov. 14, 1972). There were new code words: "selling a bit of the farm" became a commonplace phrase for economic nationalism—and more. The fear of China that had generated much public debate in the 1960s was replaced by fear of Japanese investment. Though the prime minister might have denied the charge, some saw in these kinds of statements a continuing equation between nationalism and racism.

As late as 1984, those who lamented the fated decline of British blood and customs in Australia found an articulate voice in Geoffrey Blainey, an otherwise distinguished university professor. The most generous interpretation of his position was that he merely wanted to reduce Asian immigration. Others saw in his platform a desire to maintain "White Australia" (Alomes 1988, p. 320).

The xenophobia found in the White Australia policy and its mutant forms was an insufferable nationalism writ large, and never was this more true than when Australians contemplated the north, where geography and demography combined to convince them that an invasion by the "yellow hordes" was inevitable. Thus there was not only an imperative to defend this vast northland, but also to populate it with white Australians.

In a fundamental sense, it was not a matter of how much it would cost the national and state governments to put "the right kind of people" here and there in the north; it was only a question of when and how and by what rationale. Since before the turn of the century, cattle ranching and occasional mining activities had been the dominant contributors to the economy in the north. But the needs of cattle and the nature of the land and climate were such that neither endeavor held any promise for supporting large numbers of people. Only agriculture, and not all kinds of agriculture, held out that possibility. It was merely a matter of time and of better understanding the possibilities of the physical geography of the region before irrigated agriculture would be pursued as panacea in the Ord Valley, adjacent to Newry Station.

Lately, I've been encouraging Nancy to think seriously about returning to Australia next year to trap two populations of blackhearts—yellowbeaks and redbeaks—and take them back to Irvine. It'll mean some hard hustling in Darwin before we leave, to see just how much paperwork we'll have to do—if it's even feasible, as the Aussies are reluctant to let anyone export birds. I think Nancy is keen on the idea, but she needs to think about how many birds in each population she'll need to successfully carry out lab experiments. And how many we think those at the Conservation Commission and in Canberra (the federal capital) will allow us to take. Several hundred would be ideal to get plenty of phenotypic variation and to allow for the losses we might well incur in moving them such a long distance. Without Heidi's support, I'm sure our chances of legally trapping birds to take home are virtually nil.

We've organized another trapping trip with Heidi, but have run into some technical problems. The most serious is that we don't have room in our van to transport the whole team plus field gear, and Heidi won't take anyone in her van for "insurance" reasons, she says. It's a problem that applies only to our Americans, not to her American or Australian field workers, she adds. Be that as it may, the trip is sufficiently important that one student will have to stay behind. It will be Nat, because the data she collects can be obtained by anyone, and she's the slowest worker.

I asked Heidi if Nat could stay at Keep River rather than be stuck at the hotel by herself. There's plenty of room at the park (I and others have stayed on the screen-enclosed patio without incident), and Heidi will benefit directly from the data we gather on the trip. Her response to my request was, "Talk to Keith about the matter."

I talked to Keith. I told him that we only wanted Nat to stay at Keep River for a couple of days, and that our data gathering would help Heidi and therefore the Conservation Commission.

He said, "You can't leave the students at Keep River."

"Is there a reason?" I said, puzzled by his response.

"You've been uncooperative in the past."

I waited for an explanation. When he didn't elaborate, I said, "What are you referring to?"

He walked away without saying a word.

I've no idea what this is all about. Could Terry be behind this? Might he have told Keith that he doesn't like working with us—the spat over Nat and her go-slow, let's-take-lots-of-pictures work habits? Or is all this rooted in Terry's continuing and heated battles with Neils, a relationship (to hear Neils tell it) that had gone at least as sour as ours with our students? But why would Terry think that we have sided with Neils? Is it because of his distrust and dislike of foreigners, especially Yanks?

Terry, though, may not be entirely to blame for the deeply strained relationship with Neils, who's obsessively preoccupied these days with getting yet more film on the Gouldians. He seems increasingly inclined to pursue his filming project at Terry's expense, and without regard to the scheduling of their joint field activities. Until Neils's unwillingness to help Nancy, he'd become brashly predatory with her and me, wanting to talk with us about finch behavior at the most inopportune moments.

Or perhaps the reason is the students, especially Jean. On several occasions she's given the distinct impression that her interest in Terry is more than that of a mere friend. From small things Terry says, it's clear that Jean and Nat have opened up to him about their problems with us. I have no reason to believe that Terry's made any effort to put their gripes in perspective. Given how much he and Neils have complained about Heidi's treatment of them, it would be remarkable if he didn't find it easy to side with our students—and then to make a case against us for Keith's benefit.

But then, who knows what's going on? I only have the smallest window on the many intrigues swirling around us.

We were unwilling to leave Nat alone at the hotel for several days, so I called Heidi to cancel the trip. Upon receiving the news, she reconsidered on the spot and suddenly decided that insurance wasn't a problem after all.

When Nat heard that she was going and that Terry would be riding with Heidi, she was anxious to be with him. I gave the okay, and Nat was ecstatic. She raced to tell the others the news. Jean was obviously miffed that she didn't get to ride with Terry, which only increased Nat's glee.

It's the middle of the Dry, but it rained and we were unprepared. We scrambled to protect birds in cloth bags, data sheets, Nancy's very expensive paper Munsell chips. The rain made everyone slightly giddy. Terry slipped and fell in the springs where we were trapping. When he came up, covered from head to foot in thick dark muck, he gave true meaning to the long-named trap site: Mud Springs. Presently, he disappeared. When he returned, he was hooting raucously and strutting like a rooster. He'd covered his entire body with broad leaves, which firmly stuck in the gooey mud all over him. Jean and Nat and Amy had a good long laugh. Tim, caught up in the moment, jumped in the mud and rolled around, then quickly copied Terry's attire and repeated his antic. Now, to the full-blown delight of the females, we had two showy young men displaying their best in the middle of a mud lek!

I made another trip into the farmlands of the Ord Valley, this time to learn more about how farmers control unwanted species.

Besides the long-haired rat (*Rattus villosissimus*), which can cause substantial damage to ripe rock melons, maize, and chickpeas, the land animal that has principally concerned Ord farmers is the agile wallaby (*Macropus agilis*). Among the most common macropods in tropical coastal Australia, it's been considered a pest by farmers not only in the Ord Valley but elsewhere in the Northern Territory as well, and in Queensland (in the sugarcane fields). As a control measure, the wallabies have been poisoned and shot. In the early 1980s in the lower reaches of the Ord River, pet-meat shooters killed more than five thousand (Department of Regional Development and the North West 1985).

Local banana farmers claim that wallaby numbers are increasing, and that the animals exist in "plague proportions." They like to chew bunches of bananas at night, but farmers claim they also eat melons, sorghum, and corn. Electric fences to keep them away from crops apparently are ineffective. To appease the farmers—as much as anything to permit them to vent their frustration over crop losses—the Department of Conservation and Land Management, under a "legitimate damage" provision, issues permits that allow them to shoot three or four wallabies at a time. Some local scientists claim that farmers overstate their case, that the damage caused by wallabies is minimal. Their preferred diet is grasses, though they'll eat some maize and young wheat and other emerging crops when green natural food is relatively unavailable. Wallabies stay away from recently irrigated crops because they don't like to get wet.

I take Nancy and Tim into Chloe to trap for several hours. Then Nat and I work the trap at Tanya.

Later Nancy tells me that after we left, Tim was very quiet, avoiding eye contact, not paying attention to the work of setting up the V-nets. When the nets were finally in place, Nancy asked what was wrong.

"Amy's mad because I can't leave here in September and go traveling with her. I just don't know what to do. The way I see it, I made a commitment to you to stay, but I made one to her to go. Can you see it my way? I really want to go with her, and I think you should let me go."

"Why is that?"

"Because I made a commitment to her. She has done a good job here and deserves a vacation. She's tired of taking care of Cole."

"Tim, you shouldn't have made that commitment to Amy because you had a prior commitment to Rich and me to stay as long into the fall as we needed you. That was a very explicit agreement."

"It wasn't in writing like the rest of the contract."

"That's because I didn't think I needed to draft a special contract just for you. In fact, if the tables had been turned, I would have found it an affront to ask for one under the circumstances. I had helped you out of a fairly big bind with Professor Williams. I gave you a second chance, no questions asked. I gave you research money to continue your bat work in Costa Rica, and I took time away from my own work to advise you on that project. I brought you into my lab and taught you data-gathering techniques to help you there. I have never asked anything from you in return, except for you to treat me well. Didn't we have an understanding, Tim?"

"Yes."

"And didn't Amy's desire to travel come after we agreed to pay her for services that she had volunteered in order to be with you?"

"Yes, that's true."

"So isn't it rather outrageous for both of you to expect me to let you leave early to go with her?"

"I don't understand what you mean."

"I mean, it's like slapping me in the face."

"I don't see it that way at all. Life is complex. Things change. You have to make compromises. In the spring I didn't think I would have any commitments in September. Now I do. It's that simple. If you put yourself in my shoes, you'll understand. Look at it from my perspective."

"Tim, this time I really don't care about your perspective. Rich and I cannot make you stay here or fulfill your obligation. Like all of the students, you can leave anytime. You have your plane ticket, your money, your passport, your possessions.

But if you go, don't come back to me. For help on research, for letters of recommendation, for anything."

Nancy says that a look of surprise bordering on shock came to Tim's face. "That's *not* fair," he said. "You're being really unfair. That's blackmail. A professor is not supposed to act like that. A professor is supposed to be supportive."

Nancy said nothing more. Tim stomped off, reappearing only when I arrived with the van and honked for him several times. Nancy said little on the way back to the hotel. Tim looked out the window silently. Only Nat was chatty, oblivious to the tension in the vehicle.

Nat, Jean, Tim, and I drive south, to a site where Nancy and I just last week trapped along a small creek that crosses the dirt highway running parallel to the border with Western Australia. Using mistnets that we lined up along the reedy and muddy creek, Nancy and I managed to catch several kinds of finches, including some thirty zebras. About a quarter of them had black beaks. This was a sure sign that they were young birds, and that breeding activity was probably taking place somewhere in the vicinity. Or that it had been in recent months. We hadn't had time to scout the area for nests, but with all the prickly acacia we'd seen, which zebra finches like for its protection from predators, we thought that here at last—and only a couple of hours from Newry—we might find a breeding colony of the sort we'd long sought. So we decided that it would be worth the effort if I took the team and made a daylong journey to the area to see how many active nests we could find.

Upon arrival at the site, I draw a map and subdivide it into regions. I then ask the students to choose which of the regions I've identified they'd like to explore systematically in search of nests. I ask them to mark trees and record all nests found, both those that are active and those that are inactive.

We spend all morning searching on foot in every direction for the nests. I climb a high ridge and explore the steep flanks on either side. The students take the flatter terrain to the east. We find only two nests total, and they're old.

Unwilling to give up, and on a hunch, I drive south to Spring Creek Station. After exchanging pleasantries with the station manager and his wife, I get directions to a camp a little more than twenty kilometers to the north. The manager claims that the area has "lots of zebra finches." But as I talk on about finches in general and then circle back to descriptions of the birds, I have doubts that the manager can distinguish one finch from another.

Again we search on foot, this time among two good-sized patches, each containing a hundred or more prickly acacia trees. For several hours of effort, we find only three nests with eggs. Another two nests look as though birds have recently fledged in them.

With some light left, I hurriedly drive back toward Newry. Before reaching the Victoria Highway, I take a familiar turn to the east onto Rosewood Station in search of a fenced-off bore surrounded by a score of acacias, a site where Nancy and Heidi and I had trapped several kinds of finches.

The still-familiar pond is almost completely clogged with algae. Cattle have been eating anything that grows for a kilometer or more around it. The prickly acacia inside the fenced-off water is thriving. But we see no finches of any sort. Nor, in our brief search among the trees, do we find any nests, old or new.

By this time we're all tired and, except for Nat, have slowly retreated inward in search of silence. Nat, more or less as she's done all day long, complains. Earlier, she hadn't been sure she could search for nests because there were cows and a few retiring-looking bulls where we stopped. They might trample her to death, she'd claimed. Now, despite the cool temperature and the coming of darkness, it's too hot in the van for her. She shudders at the thought that the dust circulating in the air inside the cab will have a serious and long-term effect on her sinuses.

She moans on and on, until finally I tell her that I'm fed up with her constant carping. I remind her how often before coming she'd advertised herself as a seasoned field-worker, able to cope with whatever came her way. In truth, we've discovered to our chagrin, Nat requires more attention than Cole.

Nat momentarily buckles under my harsh words. Not five minutes later, however, she picks up where she left off, announcing that I just don't understand what devastating effects the dust is having on her breathing. I'm going to be personally responsible for her premature death, she proclaims.

Unable to find words for my journal at the table in the ringers' dining area, I walk down to the hotel to get a cup of coffee and spend some time with Cole. I find him playing by himself near the door leading into the shower. Amy has just finished doing the dishes and sweeping the floor. She begins talking about Cole, says he reminds her of the brother of one of her sorority sisters. I ask her what it had been like belonging to a sorority, being president.

She says that in her senior year there were 157 girls in her sorority, Chi Omega. About a third of them lived in the house.

"What did you get from the experience?" I inquire.

"I guess I mostly learned about the range of people you meet in life. I never would have met a lot of different kinds of people if I hadn't been in a sorority."

"What kinds of values did Chi Omega promote?"

"What do you mean?"

"When you were president, what did you consider important? Was there anything you wanted your sisters to take away from the experience?"

"We didn't want Chi Omega girls to give us a bad name. We didn't want them sleeping around."

"But some did?"

"Yes, but I didn't know about it."

"Didn't know, or pretended you didn't know?"

"I didn't *want* to know."

"Being discreet was what mattered, then?"

"Sure!"

"What about other values?"

The question seems to confuse her, and when she adds nothing I let the conversation die. I get down on the cement floor with Cole and pile blocks one on top of another so that he can knock them over.

While Heidi is bleeding a Gouldian finch, she offhandedly remarks that she's been to the doctor recently and had several skin cancers removed. She has to go frequently to keep up with the rate at which they're now appearing. Her arms and hands, even her face, are splotchy, and it's common to see several red sores somewhere on her limbs, sores that persist for weeks at a time. In the harsh light of day, she often looks as if she's in her midfifties or older, rather than her chronological age of forty-four. She's long been disdainful of the sun and rarely uses sunblock.

After Heidi leaves, Jean expresses dismay at Heidi's cavalier attitude toward her cancers, her joking reference to all doctors as quacks. "I just can't understand how she can be that way," she says.

Two days ago, Nat let slip that Jean has already made travel plans to leave the country, and earlier than agreed in the contract she made with us. So far neither Nancy nor I has said a word to Jean. But Nat must have told Tim what she revealed to us, and he in turn went to Jean with the information that we know about her latest deviousness. At dinner her guilt was written all over her face, what we saw of it when she wasn't looking down or away or fumbling with her food. As soon as she finished eating, she got up and left without saying a word and went to her tent. We didn't see her for the rest of the evening.

I had a thought: Maybe the students came to Australia with unbounded images of a magical and enchanting landscape, a picture born of distance and selective memory, of having pined over too many travel-guide photographs. All of them are disappointed, greatly so, and now and then I get the distinct impression that they blame us for the monotony of these cattle-worn lands, the ubiquity of the eucalypts, the paucity of charismatic megafauna.

Thanks, I think, at least in part to the eruption of Mount Pinatubo, the sunsets here have been spectacular. How seldom we've found time to appreciate them! The sunrises have been accompanied by a glorious litany of birdsong, unparalleled in North America. To the students, though, getting up early has only meant a day of work that ends by early afternoon. On days when they haven't had to go into the field, they sleep late, the little corellas no more than an irritating alarm clock they seem fully able to ignore.

The whole pattern of trapping has changed. At Dingo Creek we caught fourteen masked finches and six Gouldians and no longtails. For the first time, a small kangaroo came to water. He tore through one of the nets.

Amy has set herself some new goals: getting Cole to kiss her on the lips and to give her a high five. She's sewing a binocular case for Tim, similar to the one Nancy made for herself some years ago.

Once again, Amy's been throwing up. Too much junk food and candy? Who knows? Tim's not vomiting, but he says he doesn't feel well. One minute he blames it on some "rotten cabbage" we all ate, the next on the water we drink. He's said no more about leaving early with Amy.

Although Nancy and I had planned to keep Nat with us into the month of September, perhaps until the second or third week, I now think we'd do better to give her a little extra money and ask her to leave when Jean does. I'm exhausted from her nonstop complaints, her poor field skills, the constant need to prod her to hurry along, her unreliable observational abilities. I've said nothing to Nat in recent weeks about the DNA fingerprinting. Perhaps, I now think, her reneging on her promise is a great blessing in disguise. It'll be one less worry for Nancy to deal with before dismantling her lab for the move to California.

Nat's mail contained very bad news. The father of her former boyfriend, the Wittgenstein wonder, died. She's been walking around in circles, saying out loud, "He died so fast." He died three weeks after being diagnosed with cancer. She spent the whole morning writing a letter to the widow.

Jean is leaving tomorrow, and we've definitely decided that Nat should also go. Both of us have had all we can take of her. We also see few ways to use her, now that we've decided to shut down the trapping. We'll take Tim and Amy with us on a long trip to Broome and the west coast of Western Australia to see how much longtail beak color changes. And if, by chance, we finally happen upon some breeding zebra finches . . . well, what will we do?

As soon as Jean is on her way, I'll have to take Amy into the hospital. Suddenly she's worried sick that some spots on one breast and one on a thigh are skin cancer. She's not shown either of us the spots. I suspect it's Heidi's skin cancers that have scared Amy, perhaps finally made her think hard about her long afternoons fully exposed, not a drop of sunblock on her body.

I've just learned why we were unexpectedly unwelcome at Keep River, why Keith accused me of being uncooperative. It turns out that Terry told him Nancy and I were working only with blackhearts and didn't give a damn about the Gouldians! We are alleged to have ignored them. In talking with Heidi, I learned that she too sent a similar message to Keith. Little wonder that Keith found us "uncooperative"—aggressive imperialistic Yanks once again ripping off helpless Aussies.

I cannot fathom why we have been portrayed this way. Heidi's database on Gouldians has been greatly enlarged because of us. Nancy has openly shared insights and hypotheses with Heidi, shown her new methods of analysis, alerted her to problems with the Gouldians. These revelations are as big a puzzle as the students' behavior.

When I told Nancy how we had been represented to Keith, she was furious.

We took everyone into Kununurra for a going-away dinner for Jean. I bought her a nice card and flattered her with some words about her field prowess. I made every effort to show as kind a face as possible to someone I've come to dislike intensely.

We all had honeyed prawns, and I bought a couple of bottles of special wine. Hard as Nancy and I tried to give Jean a happy send-off, the mood was somber. Jean seemed to find it difficult to smile. I think she knows how we really feel.

On the morning of Jean's departure, Nancy and I told Nat that we'd like her to leave three weeks earlier than originally planned. We said we'd pay her for this period, since we'd made a commitment to her. When she asked why we weren't going to keep her for three more weeks, we said that trapping on Newry had slowed to a trickle, and that Tim and Amy were going to Timor for a week on a brief vacation, while Nancy and Cole flew to Irvine to sign papers on the house we'd bought sight unseen. I'd be on the road again, in Queensland, trying to get information on the eastern geographic range of red-beaked longtails. Nat said she'd love to go to Queensland with me, explore new terrain. I said, No, that's impossible; I'm going alone. She cried. She promised not to complain ever again if I took her with me.

Since we broke the news to Nat about her early departure, she's returned to Nancy and me three times to ask why she can't come with me.

Later I drove Jean into town so she could catch her bus. On the way, I asked her if she wanted to say goodbye to Terry. Tim had told me that Terry was staying at the Desert Inn in town, and I could drop her there.

Jean said, "No, he's not there."

I then told her I needed to stop at the Desert Inn to pick up a shirt I'd left behind when I stayed in town for a night after snooping around all day among Ord Valley farmers. In the lobby I ran into Terry. He said he was expecting Jean, that she knew he was coming.

Fool that I am, I was surprised to hear it. Why was Jean bothering to be deceptive? A habit she just can't break?

At the bus stop I helped Jean with her luggage and stood around trying to think of something nice to say—wondering what other deceptions and unwelcome surprises lay ahead, and what was being said about Nancy and me as Jean cuddled up to Terry. I walked away with nary a final goodbye and sped down the street onto the Victoria Highway to return to Newry. Within minutes I felt an immense sense of relief.

Now that the cattle have been rounded up, branded, and dehorned, and some shipped live to Indonesia, it's quiet around Newry at night. No more mooing and plaintive wailing, no heavy bodies banging against one another and the unyielding iron. Only the constant, pounding hum of the station generators reminds us of the isolation and why people live here.

Wes, alone among the ringers, is still around. He's mending fences and bores, doing odd repairs on the station's old and broken equipment. His work is a reminder not of prideful ownership, but of an absentee landlord whose sole abiding interest is extracting as much as humanly possible from this meager habitat.

Nancy and I tell Nat that we'd like to give the same going-away dinner party for her that we did for Jean.

She says, "I can't do that, I'm going to have dinner with Terry and Neils the night before I leave."

"Okay, whatever makes you happy," I reply.

The next morning, her eyes filled with tears, Nat said she didn't have plans for dinner with Terry and Neils after all. It was no more than a wish. Now she didn't know whom she wanted to have dinner with on her last night in the area.

An hour later Nat came to me and said that the "mixup" over her final dinner was all my fault!

Nancy and I sit and tally up the final banding and recapture data on finches (see Table 2).

In spite of the many problems, we have managed to collect an enormous amount of original data. In this sense I think our field season has been a resounding success. Some of the data can be worked up and analyzed and sent off for publication after Nancy completes her move to Irvine and settles in to a normal routine. Some

Table 2. Tally of all finches trapped during the 1991 field season.

Species	Newry sites		Elsewhere		Total	
	b	r	b	r	b	r
Longtail	1,425	779	499	35	1,924	814
Masked	357	103	161		518	103
Gouldian	217	48	34	1	251	49
Zebra	44		35		79	
Pictorella	6		7		13	
Double-bar	28	1	27		55	1
Star			75		75	
Crimson			35		35	
Total	2,077	931	873	36	2,950	967

Note: b = banded; r = recaptured; each recaptured bird is counted only once, even though some birds were captured numerous times.

of the data will, for some time, serve the primary purpose of background information, which will prove useful in designing laboratory experiments. Some of the material—much of it—will be the source of hypotheses, to be tested in the lab and possibly in the field should we or some of her graduate students return.

The effort spent on mapping nests and characterizing the cavities chosen for use as opposed to those not colonized has convinced Nancy that Newry longtails are in no way nest-site limited. These birds use cavities less than a meter off the ground and as high as eight or nine meters. They use cavities facing in every compass direction, those with large entrances as well as small, those with horizontal shallow cavities, and those with deep cavities angled toward the tree's roots. We have estimated that fewer than one in ten suitable cavities has been in use at all during the season, and at any one time the active breeders might use 3 percent of the available cavity sites. We have seen almost no signs of overt competition for nest sites.

This type of information is very useful to Nancy in developing assumptions and lab experiments. In some groups of birds, for example, bib size is a badge of social dominance that may lead to priority of access to resources, including food, nest sites, and territories. Because the principal food, grass seed, is widely scattered, and because feeding finches are quite vulnerable to predation, finches in feeding flocks show relatively little aggression. Thus, if bib size is a badge of dominance, Nancy reasons that it should confer priority of access to preferred nest sites. If we're able to import blackhearts and get them established in Nancy's lab, she can run mate-choice trials to determine if Newry females—or their immediate descendants—prefer large-bibbed, long-tailed males. Although she has never been able to locate a plumage-based mate preference in zebra finches, she expects she would find such a preference in longtails. She will probably also run nest-box trials to determine if large-bibbed (or long-tailed) birds have priority of access to nests. She does not know what to expect here, but she will not be troubled by negative results, because the field data indicate lack of direct competition for nests.

Trapping data through a considerable portion of the breeding range will allow Nancy to characterize the changing phenotype of the longtail throughout the Northern Territory and eastern Western Australia. They also will allow her to decide unequivocally if the populations she would like to study in the lab are those we have come to refer to as Larrimah and Newry populations. It appears that there might be a distinct trade-off in effort spent on various signals. Birds from Larrimah have the brightest, reddest beaks, but also the shortest tails and smallest bibs. By contrast, Newry birds have the biggest bibs, the longest tails, and the yellowest beaks of the major populations we have sampled. What is happening at Newry? Nancy wonders. These birds have beaks more yellow than those captured farther west. They also have the greatest apparent sexual dimorphism of all the populations examined.

The intensive trapping information at Newry has revealed several things. First of all, there is a lot of turnover, and it takes a great deal of effort to recapture birds. This is a bad omen for future fieldwork. Still, Nancy is anxious to return next year, if possible, to determine what percentage of the many birds we banded at Newry can be recaptured. She expects the percentage will be low. Second, we have ascertained that the vast majority of (if not all) longtails leave the breeding grounds at Newry before the end of the Dry. As Dingo Creek was drying up, the recapture rate plummeted; it was no different at the artificial waterers at Chloe and Tanya. The birds we captured early and frequently in the first month of work (the "local residents," as Nancy has tentatively dubbed them) disappeared well before the last wave of birds arrived. Perhaps these individuals reproduced early and left with their young before doing so got "risky."

These results indicate that there really is a lot of movement of these small birds at the breeding grounds, and that they are far from sedentary. Nancy finds the small-scale phenotypic radiation of the long-tailed finch even more remarkable, based on these findings.

We have not obtained as many behavioral data on breeding longtails as Nancy would have liked. No doubt this reflects, in part, our observational skills and the time devoted to this activity. The birds were definitely very shy around the nest, a sign that these finches are not protected from nest predation by virtue of nesting in cavities. High rates of nest failure and the spatial patterning of nests are consistent with this inference. We did get information about the rhythm of the nest cycle, the pair's duet. Nancy said that the few reports she has read claiming that the pair bond is relatively tight in this species are true, at least if we base pair-bond strength on the amount of time mated birds spend in close proximity. And what a contrast to the Gouldian finches, whose males and females are seldom seen together at the nest!

We have blood samples from longtails and other finch species. Nancy's not sure if there are enough for meaningful population estimates of extra-pair fertilization in longtails. It might be better to use the blood to examine finch radiation at a molecular level. We succeeded in getting asymmetry measures on enough Newry adults that Nancy will have a chance to make something of this data set.

The data we have gathered do not provide any direct insights into the decline of Gouldian finches, Heidi's preoccupation. Based on anecdotal evidence, however, it appears that *if* "something" is limiting the number of Gouldian and long-tailed finches at Newry Station, longtails may well have a competitive advantage over Gouldians. Blackhearts seem less choosy with regard to nest sites, they seem quicker to colonize successful nest sites that have been abandoned for unknown reasons, and they seem to have behavioral traits while nest building and guarding their young that may make them somewhat less subject to predation than Gouldians. These are the kinds of inferences with which Nancy—short of having many more data—does not feel comfortable, and would make no attempt to publish or broadcast among her colleagues. I, however, believe in a different epistemological reality than Nancy, and I am quite content with these kinds of tentative conclusions. I plan to hold fast to them until another field-worker produces convincing data to the contrary.

How much more pleasant our days and nights are with Nat and Jean gone! In the few days since their departure, Nancy and I have developed a much warmer relationship with Tim and Amy. They're more open, and they seem more honest in everything they say and do.

This long journey west to Broome has yielded less than we hoped. We trapped at the Royal Australian Ornithological Union Bird Sanctuary, where some two hundred fifty different species of birds have been identified. We set up three nets. But after three days of trapping at a site alleged to be "great" for blackhearts, we caught a mere seven birds. At least they all have the same beak color—yellow—and Nancy is gratified that they're close enough to the Newry beak color for her to work on the assumption that the long geographic cline to the west of Newry is gentle, probably lacking in surprises. It's nothing like the sharp gradient between Newry and Larrimah in the Territory. Just as well, I suppose, since most of her data on longtails are from the Territory.

The only incident of note occurred a couple of nights ago when the five of us—Cole, as always, goes everywhere with us—went to see the movie *The Doors*. When we returned to the van, we discovered that a side window had been completely shattered by a large rock. A restaurant waiter said that an Aboriginal couple had gotten into a loud argument and that the woman had started beating up on her husband or boyfriend. Not satisfied with the results, she picked up several rocks and began throwing them at him. The van window got between her anger and the person she wanted to hit. I filed a complaint at the police station, but was assured that the likelihood of recovering any money or of the woman's being prosecuted was slight.

Tim, on the return trip, talked about classes he had taken as an undergraduate, and at one point noted that a philosophy professor had remarked to him, "Tim, you are the most untroubled person I know—and, alas, therefore the most unphilosophical." Tim did not verbally reflect on the insight that now strikes me as a profoundly revealing truth.

When we get back to the Blackheart Hotel, a fax is waiting for Nancy. It reads: "Dear Dr. Nancy Burley. I will not do the DNA fingerprinting. Nat."

Amy and Tim have left, and Nancy and I have begun cleaning up, packing the tents and other supplies, deciding what to sell in town at a secondhand store, what to send to Irvine, what to pack inside the van that Alan has told me we can leave indefinitely under the shed to the west of the station generators. We have made tentative plans to return next year if we can get the permits in Darwin to legally export both red-beaked and yellow-beaked longtails.

Nancy spent several hours cleaning the refrigerator while I scrubbed here and there and gathered trash that we'd accumulated and not taken to the dump area behind the horse corrals. It'll take another couple of days before we're ready for the trip to Darwin, and then home to Illinois.

Nancy calls Dave, her graduate student, and he has nothing but bad news. No, *terrible* news is more accurate. Jean returned to Illinois before Nat, and she delivered a note from Nat to the director of the School of Life Sciences complaining of poor treatment by us. What is Nat referring to? My outburst reminding her of her moral obligation to do the DNA fingerprinting in exchange for the privilege of working on an M.S. degree and coming to Australia with us? That I made no attempt to hide my anger when all of them save Amy made Cole's first birthday party memorable in ways I don't want now, or ever, to remember? Perhaps she complained that I had no right to tell her to quit kissing and fondling birds? Omitting, no doubt, that when I did so we had dozens of birds at our feet in calico bags and my concern was that we not kill or injure one of them—and never mind her sensitive female nature.

Dave also informs us that the director has thrown out the students working in Nancy's finch lab and shut it down. Never mind Nancy's right, and mine, to a hearing on charges against us. Never mind a research program that demands continuity. Never mind a lucrative contract with the National Science Foundation that like so many federal research grants goes a long way toward supporting graduate and undergraduate education at the University of Illinois.

To make matters worse, we also learn that the chair of Nancy's department has stood on the sidelines and done nothing to prevent these kangaroo-court proceedings. He has not made the slightest effort to contact us and inform us of the charges, or ask what happened, or give any rationale for allowing the director to sabotage Nancy's research.

There's more to this surreal lynching from afar, Dave tells Nancy. Our dear friend Lana Ostinger—Nancy's colleague in plant biology, our frequent dinner guest along with her mother, the one person other than Nancy's chair who could easily get to the director and tell him to stop the proceedings—has disowned us just as completely as Nancy's chair. She can't be bothered to "get involved."

It was early afternoon when Nancy walked back from the Rec Room and told me about the charges against us. When she was through, I sat down on the concrete lip of the old Center Camp building that Nancy and I had talked about using to house longtails if we come next year, and I fell silent. We held each other. I felt spent, and numb all over.

What are we to say or do? We're halfway around the world. We can only imagine the seriousness of the charges against us: try to visualize what has become of Nancy's lab, and her needy finches, and the innocent students; wildly guess at how intricate a story, with all of its Nat-orchestrated fictions, is being fabricated to legitimize the "justice" meted out in our absence, without so much as a word from us or in our defense.

We have a final meeting with Heidi in Darwin: to share data, to go over what we've accomplished, to talk about where future research might go. With Heidi's assistance, we manage to arrange meetings with those at the highest levels of the Conservation Commission of the Northern Territory, and there make a case for importing into the United States seventy red-beaked and seventy yellow-beaked blackhearts, for breeding and experimental work in Nancy's still-under-construction lab at the University of California. Objections are raised about our taking any birds at all from the sites where we trapped, objections raised not because of any real or imagined impact on Gouldians or the local ecology, but because of anticipated reactions from one or two vocal environmental activists in Australia who are adamantly opposed to exportation of any Australian wildlife, regardless of the purpose. But Nancy and I persist. When we at last leave the Conservation Commission, we have in hand signed statements that will greatly facilitate more formal requests by Nancy to trap blackhearts next year and return home with them.

6

Illinois Aftermath

The flights from Kununurra to Darwin to Sydney to San Francisco to Chicago to Champaign were full of apprehension. The questions kept repeating: What exactly are Nat's charges against us? How serious are they? Did she give the letter to Jean after we asked her to leave early, or before?

Nancy read and watched Cole and I slept. And then I read and played with Cole while she slept. I walked the aisles, I engaged stewardesses in small talk, I watched mothers tend babies and old men snore and young girls play with their hair. Anything to keep my mind off the problems we'd soon have to confront.

Before leaving Newry, we'd agreed to not talk anymore about "the problem," not until we had more facts and knew what we were dealing with. And we did a good job of it until we left O'Hare Airport for the short flight to Champaign.

Nancy opened with, "I guess I made a real mistake with Jean. I just didn't want to believe what I now see clearly."

"I should have pushed you harder not to take her."

"Well, it's history now."

She fell into silence, and I did the same. But I could not get the students off my mind. I went back over this event, that happening, and I kept going—remembering everything I could. The questions kept coming. Who were these students? What were their motives? What should we have done differently? How would we—or I—plan another extended field trip, with the kind of knowledge now in hand? No single answer seemed to suffice, and yet certain thoughts kept recurring.

Contrary to my insistence that we persist with the students no matter what, I now think that we would have been wise to send all of them home. Nancy said as much fairly early on. I am the one who insisted that the students would change, that we had to give them another and yet another chance. Isn't that what we all want at some point in our lives?

At what point we should have sent them home, I'm less certain. Perhaps as early as when Nat said she would not do the DNA fingerprinting and it looked as though Jean and Tim were not about to come out of their inexplicable funk. Certainly by the time Amy announced that she no longer wanted to baby-sit and had to spend time in the field with Tim. Beyond doubt after the students returned from their Kakadu vacation and then quickly reverted to their former ways.

Once we had sent them home, we would have been forced either to recruit backpackers in Kununurra to help with the trapping and data recording, or to cut back substantially on our objectives and perhaps work more closely with Heidi and her team. The obvious risk with backpackers would have been inexperience, particularly in freeing birds from mistnets without injuring or killing them. That was of paramount concern to both Nancy and me. Further, Nancy would have had to train some of them in how to take measurements and draw blood. The logistics of picking up and dropping off the replacements would have been time-consuming. I'm not sure that Alan would have approved of several backpackers camping for a week or more on the station, and then leaving when others took their place. It might not have been wise for us, either—opening the door for theft and for some of the same interpersonal problems we had with the students.

We made a major mistake in trying to get close to the students, in treating them as responsible adults who would live up to their commitments and not take advantage of us at every opportunity. It has been tempting to see these students as representatives of a new and misguided generation, but this is most likely a mistake. The students, I remind myself, are just people. They did what most would probably have done when they saw there was little to lose by behaving badly. As for the gains from doing so, who knows how these were calculated?

Nat had to know that breaking her promise to do the DNA fingerprinting would jeopardize her relationship with Nancy, and possibly her master's degree. But it's questionable that Nat ever really wanted that degree—or, assuming she did, that she saw clearly the possible consequences of her behavior. She's unpredictable, neurotic, self-absorbed, a directionless waif, a spinner of fantastic stories small and large. People as quirky as Nat are often irrational.

Tim does seem quite rational, and he made it evident that he wants to get into one of country's best graduate programs in ecology. Usually students with this aim are eager to please, to impress, to run the extra mile. In return they want an outstanding letter of recommendation for graduate school. Tim is not required to ask Nancy for a letter, but in light of her very fine academic reputation and the value that ecologists place on field experience, he had to have seen the considerable value of not leaving a bad impression. But Tim, it seems, fell prey to Jean's whims and dictates, a person much closer to him in age and concerns than Nancy or me. Without Jean around, Tim might well have lived up to the expectations Nancy and I had of him.

I believe that Jean was the person most responsible for the sour social environment, for most of the human problems that permeated our field season. She proved uncooperative, duplicitous, and uncommonly smug in her assessment of our research mission and execution. She was the worst kind of critic: disruptively derogatory behind our backs, but utterly incapable of coming to the fore and offering constructive suggestions about whatever was proving troublesome.

Nancy, it is now more obvious than ever, had no hold whatsoever on Jean. By the time Jean joined us, she had no intention of working for an advanced degree—with Nancy or anyone else. All that was at stake was how Nancy and I would react to her behavior; for neither of us is in a position to sway those who employ her at the Illinois Natural History Survey. And Jean, like the other students, knew that Nancy had taken a job at the University of California, which all but eliminated her potential influence.

We made another miscalculation in assuming that we would not be seen as "parents," or people of an older generation who "just did not understand" and perhaps could not be trusted. The generational gap was there in the beginning, and nothing we did attenuated it. In the diffuse Illinois environment it was not that apparent; it became increasingly obvious, however, as Jean and the other students reaffirmed their values and opposed them to ours in that isolated and difficult-to-escape outback frontier.

We should have treated the four students as employees. We should have paid them so much per hour (or defined workweek) to perform specified tasks. Beyond that we should have more or less let them fend for themselves—live where they wanted to live; eat whatever and wherever they chose; and entertain themselves apart from us. We might not have gotten to know them very well, but I suspect we would have been treated with more respect, received more for our money, and endured few of the betrayals and disappointments that we encountered.

Many ecologists who regularly do fieldwork don't pay the students who work for them. They may provide food and a place to live, but no more. Fieldwork then becomes a kind of privilege, an experience to be highly valued because not every one is given the opportunity, and because that experience can be turned to advantage when applying for graduate school or work. Such an approach constitutes exploitation, and for this reason is not appealing to either Nancy or me. Yet, alas, because the practice is part of what might be called the culture of ecological fieldwork, it may—on the whole—produce better results that ours. We gave the proverbial inch by warmly embracing the students and paying them well, and they took the proverbial mile.

A major, if not *the* major puzzle, for us is why Jean behaved as she did. She alone knows the answer—and it may be ambiguous even to her. I suspect that no single influence drove her behavior, and perhaps one impetus that ruled early on gave way to another as she spent more and more time with us.

Possibly Jean's behavior was a kind of "revenge" because Nancy had so readily accepted her decision to drop out of the doctoral program. Did Jean expect Nancy to object to this decision, to extol her virtues and entreat her to reconsider? Jean's mother had once suggested to Nancy that she should "work on" getting Jean to return to the doctoral program by "wooing her," telling her what a waste of talent it was for her to leave. Nancy had explained that students without strong motivation should not be encouraged to continue in graduate school, regardless of how smart or creative they are. Jean's mother had been irritated by Nancy's response. Was the mother, in this regard, a mirror reflection of the daughter?

Jean may have thought from the beginning that Nancy and I were going to give her a major voice in what went on in the field: when and how to collect data, which data to collect, and the like. Buttressing her considerable ego were the two field seasons she had experienced as an undergraduate, her robust self-image as a fieldworker par excellence, and Nancy's reputation as an experimentalist with limited field experience secondary to her laboratory skills. If more authority is what Jean wanted, she made no effort whatsoever to convey this fact to us; when asked for input, she was mute.

It may have been that Jean was still reeling from the death of her husband, and the resulting confusion that found her in a serious relationship less than a year after his death. Alfred, the adoring boyfriend, had known Jean for a long time. Once Eddie died, Alfred not only pursued his romantic interest in her, but he apparently found an immense amount of time to give Jean what she needed most in the months after her husband's death—someone who would listen to her pine for what she had lost, and complain about how so many of Eddie's friends and acquaintances had—in her estimation—quickly forgotten him. Once in Australia, Jean no longer had Alfred to console her.

Perhaps all of these reasons overlook something even more fundamental: the predicament Jean found herself in when she had more or less committed to joining our team, yet did not formally do so until the very last moment—and then realized that her heart was not in the venture. She did not really want to go to Australia with us. She was quite involved with Alfred when I invited her to join us, and I suspect that involvement only grew in the many months preceding our departure. Jean may have felt trapped by her decision. Backing out at the last minute would have made her look unconscionably shabby in the eyes of her central Illinois peer group, and she has the kind of pride that needs to avoid this kind of embarrassment.

Once in Australia, Jean could only imagine all that was not happening, or happening adversely, in a once-budding romance with Alfred. Distance plays extravagant tricks on the imagination, particularly when the imagined losses involve matters of the heart. In the field Jean anxiously awaited Alfred's letters. She read them repeatedly in her hammock and in her tent, and—given the twisted and unpredictable ways the mind can work—she may well have come to resent Nancy and me for her time away from her boyfriend. Alfred's absence from her daily life, she may have reasoned, was our fault. But then who knows?

Once home, Nancy immediately looked to her huge populations of zebra finches. She was relieved to see that although the two research students working for her had indeed been banished from her lab, the birds had been properly cared for.

She had a couple of meetings with the chair of her department and alas, the stories were all true. He had done nothing to protect her interest, or mine; absolutely nothing to slow or halt the cumbersome university machinery that, at this time in history and almost everywhere in America, works on the twin assumptions that all student charges are legitimate and that all faculty are guilty until proven innocent. Given what we knew about the chair and his administrative skills, we couldn't have expected much. Still, it was hard to believe that he and, worse, Lana Ostinger had so completely cut us loose, done nothing to protect our most basic rights.

On our return from Australia, we decided that I would not attend any of the meetings Nancy had with her chair. I not only disliked the man, I didn't trust him.

These were feelings that long predated our trip to Australia. Nancy, knowing my willingness to engage in confrontation, thought it best to talk to him on her own.

Nancy also made an appointment with the director of the School of Life Sciences. I insisted that I wanted to attend. But Nancy raised the same objections as before.

"Take me along," I said. "Otherwise you're going to be ambushed. He's going to call in the chair and the two of them are going to beat up on you."

"It's not going to happen," she said. "I have assurances that it'll be a one-on-one meeting."

Nancy went to the meeting alone. Both the chair and the director were there. As Nancy related the episode to me, the chair looked mostly at the floor and said nothing during the entire meeting. The director was congenial. He asked Nancy to explain what had happened. She gave a detailed account of the students' behavior—the lies, the betrayals, the money stolen, the vacation we had given them, just about everything of significance she could remember.

When she had finished, she asked the director if he had any questions. He had three. Were the students always in possession of their passports and their plane tickets? Were they ever refused money by me? And were they free to leave at any time? The answers, Nancy replied, were yes, no, and yes. She then said that I gave the students money anytime they asked for it—even money they hadn't earned; that Nat had begged to stay and travel with me—and we had told her no; and that if the students had wanted to leave, it would have been easy: they were a mere hundred meters from the Victoria Highway and an effortless hitchhike west into Kununurra, or east as far as they wanted to go.

The director had no more questions. So it was Nancy's turn. "Why," she asked, "did you unilaterally shut down my laboratory?" She reminded him that the lab operation was funded by a grant from the National Science Foundation, that she was halfway around the world when he did it, and that he hadn't even had the courtesy to tell her what he had done, to say nothing of providing an explanation.

"I did it to protect the students in your lab," he said.

Taken aback, she said, "Did you know that two of the students had been working on honors projects for over a year at the time they were evicted?"

"No, I didn't know that."

"Are you aware of how upset the students are because of your action? I have talked to them since returning and they're insistent that they felt perfectly safe in my lab, no different from other students who have worked with me."

At this point, the director claimed that the parents of one of the students had lodged a complaint. Nancy was even more startled by this revelation. She asked the director if a single one of the more than two hundred fifty students who had worked in her lab over the previous twelve years had complained to anyone in the administration about their treatment in her lab.

"Not that I'm aware of."

"What was the threat that these students are alleged to have felt?"

He hesitated, looked for words, then said that some "graduate students" (whom he would not name) were telling the undergraduate students "scary" stories about what might happen to them if they stayed in the lab any longer.

"It sounds to me like these graduate students—whoever they are—were making veiled threats, and it is they who should have been disciplined."

The director shrugged his shoulders and turned away. Suddenly he became stone-faced. The conversation had obviously become uncomfortable.

Nancy wanted more answers. She said, "What, specifically, have Rich and I been charged with?"

"There are no charges, either formal or informal."

"What does Nat's alleged note or letter contain?"

"I haven't seen it. If you want to know more about it, you will have to see the ombudsman."

"Has the ombudsman interviewed Nat?"

"She has."

At this point, in view of what he had just said, Nancy insisted that he, the director, write a letter on university letterhead to the effect that no charges had been brought against either of us.

He smiled and said, "I'll be happy to do so."

Nancy's patience was gone. She pursed her lips and stared hard at the director, then at the chair. She told them that both had been sorely derelict in their responsibilities.

The chair's eyes remained glued on the floor. The director met her look and said nothing.

Nancy got up to leave, and as she pushed the chair toward the director's desk, she said that someone in the university should take the initiative to get Nat psychological counseling. She reminded the director to send her the letter he had just agreed to write.

"No problem," he said, getting halfway out of his chair. At which point he peremptorily dismissed the chair, who seemed unusually anxious to leave.

Eager to get away from the two administrators and this unpleasant experience, Nancy hurriedly walked down the narrow, windowless corridor of a lifeless research building constructed in the 1950s that one of her colleagues often referred to as Eisenhower bizarre. Whatever her anxieties, the situation was apparently even more stressful for her normally loquacious, always-eager-to-hug chair. Now he was uncommonly silent and, quick as his pace normally was, he stayed a measured two or three steps behind her.

At the first opportunity, Nancy veered into a seldom-used hallway to get away from him. As she did so, she abruptly stopped, looked back, and said "Goodbye, Lewis." There was a finality in her voice that he could not have missed. Lewis didn't lift his eyes or respond. Nancy wanted nothing more do with him—ever. She would not leave the department and the university for another two months.

By the time she reached the front door of the dingy home we were renting, Nancy realized that Lewis, her chair, had been one of the gossipy "graduate students" to whom the director had been referring. Perhaps, in fact, he had been the *only* "graduate student" who had allegedly told Nancy's honor students scary stories about their fate if they stayed in the lab. She also reached the tentative conclusion that the ombudsman had questioned Nat enough to understand that she was dealing with an unreliable witness to history. Her mind still on full throttle and now stitching together a mosaic of quotidian facts, Nancy realized that Nat had written the complaining note *before we* refused to allow her to travel with me in Queensland in Nancy's absence. Nancy could not help but wonder: was Nat planning to claim that Rich had raped or kidnapped her? Absurd in Nat's fantastic world? Hardly.

We decided no worthwhile purpose would be served by either of us talking with an ombudsman who had previously brought two of Nancy's undergraduate students to tears when they innocently tried to help a friend who had been raped.

I buried myself in writing, in a university office to which only Nancy and I and the janitors had keys. I spent long hours in the library preparing a course on conservation that I'd be teaching at the University of California shortly after our arrival.

One day Nancy ran into a colleague of sorts in the Natural History Survey, the very person who had been Nat's adviser when she searched for cowbird eggs in the Shawnee National Forest. He opened with, "Why didn't you ask me about Nat before taking her to Australia? I would have told you she's a nut."

"I did!" Nancy replied. "You said nothing, only that she was very good in the field. And great at finding cowbird eggs."

Two weeks later Nancy learned that this same colleague had just accepted Nat as a graduate student. Somehow the failing grades that Nancy had given Nat, and the Australian experiences that she and I had shared with various people, meant nothing. We were enmeshed in a moral order that left us dumbfounded.

Nancy did not receive the promised letter from the director of the School of Life Sciences. Repeated reminders sent to him and his secretary were ignored. But in her final meeting with him, to settle financial accounts and have the rest of her NSF grant transferred to the University of California at Irvine, Nancy discovered that our misbegotten adventure had yet one more twist. An unthinking accountant had paid Jean twice for her time with us. Jean, with months to catch the error, had said not a word about it to anyone. Given Jean's fussiness about regularly checking my bookkeeping while in the field with us, there could be no doubt that she was aware that she'd cheated Nancy—and for a substantial sum of money. I could only conclude that Jean—the field assistant "from hell," to borrow one of her favorite expressions—had a bottomless bag of machiavellian tricks.

With a full U-Haul van behind us, we left Champaign forever in the late afternoon of one of the coldest and snowiest December days I could remember in the nearly twelve years we'd spent in central Illinois. In three hours we skidded and crawled all of forty miles. With nightfall upon us, and only the truly insane still on the highway heading south, we checked into a motel and put Cole and Dakwa Waza Garou, our African gray parrot, to bed. Nancy and I went out to the trailer and opened the rear doors and for a long, silent moment stared at what we'd feared. Shivering as much from what we saw and the road behind us as from the cold, we dumped into a giant trash bin all the plants so carefully nourished in our Illinois home for so many years—now totally frozen.

Postscript

Soon after our arrival in California in early December 1991, Nancy began applying for the variety of permits needed legally to export one hundred forty longtailed finches from the Northern Territory and to import them into the United States. In June 1992 I flew to Darwin and took a bus to Newry Station, where I picked up the van and supplies we had used during the previous season.

When Nancy and Cole arrived, we immediately began driving on accessible tracks at Newry to identify a number of trap sites from which we would take a combined total of seventy yellow-beaked longtails. Several weeks later we carried out a similar exercise around Larrimah and on the cattle stations to the south, where we would capture another seventy longtails, all with red beaks.

The three principal trap sites where we had caught most of our birds in 1991 were largely bereft of finches. Dingo Creek, one of our most productive sites the previous year, still had water in the pool. But there were few finches, and we got none of our take-home birds there. The situation was even more bleak at Chloe and Tanya. We saw no finches on the days we were there. Heidi had stopped trucking in water to attract the Gouldian finches, and without water few birds of any kind were to be seen in the trees with which we'd become so familiar.

Alan, the station manager at Newry, once again invited us to live at Center Camp. But the hotel was nothing like I remembered it after Nancy and I had scrubbed it down and wiped clean the refrigerator and taken away all the garbage. It was dirty, grungy, trashed. Much like the first time I'd seen the small metal building, strewn hither and yon inside were beer cans, hash and bean cans, dirty rags, old newspapers. Termites had moved in and built a long rectangular mound several millimeters high, diagonally across the cement floor where Cole had played and bathed in the tiny pink tub. The green shower curtain we'd left behind was still there, now caked with dust and much uglier than I'd remembered.

At the sight of all this, and when negative memories of the previous field season began rushing into our minds, we decided not to camp anywhere near the hotel. So for several weeks during our stay, the three of us slept on the floor in our double swag in the tiny cook's quarters that adjoins the station-hand kitchen. We had no table, no chairs, only natural light or our flashlights. All the birds that we captured on Newry, as well as all of our supplies, were piled at our feet or along the walls.

Range conditions at Newry were ruinous. There had been little rain during the previous Wet. Wes, still the station's head ringer, opined that by the end of the Dry in late September or October the cattle would be in serious trouble. As evidenced by the abundance of road kills on the highway west to Kununurra, the station cattle were proving to be desperately hungry as early as July.

Kim, the cheery and reliable Aboriginal ringer from Halls Creek, had returned for another mustering season. Sandy, the apprentice jackaroo who had driven the station roads as if they were superhighways, was now in a technical school in New South Wales. He was learning how to be a cotton farmer. Wes said that Sandy had all the mechanical skills necessary to be an all-around ringer, but no understanding of animals.

In addition to our residence at Newry, we had two other bases of operation. One was a long cramped trailer in a trailer park along Lake Kununurra, just across the border into Western Australia. Some of our Newry trapping was conducted from this site. We couldn't stay as long as we would have liked because of restrictions on bringing Territory birds across the border into Western Australia. Our third base of operations, used for trapping all the red-beaked longtails, was a rented trailer in a trailer park in the roadside settlement of Larrimah. The trailer was comfortable for us and was also a much better temporary home for the finches than the one we had in Kununurra.

Even though Cole charged through more than one mistnet and on several occasions managed to get dirtier than a messy truck mechanic—and had a sixth-sense inclination to wander off just when the nets were full of birds—he was the not the problem we'd feared without a baby-sitter. More difficult was finding enough decent trap sites, as required by The Northern Territory government. (By the terms of our trapping and export agreement, we were not allowed to take more than ten finches from any one location.)

Twice on our trapping forays I let slip to inquiring strangers that we had permission from the federal government and from the Conservation Commission of the Northern Territory to return to the United States with some of the birds we were catching. This was a mistake, for it rendered the few Australians to whom I told our intentions singularly possessive. Worse, they couldn't seem to understand how it was possible that their government would allow any foreigner, biologist or not, to take *their* native birds to America.

About halfway through the field stint, and with about half of the birds to take home in hand, the engine block on the van cracked. With no options, and at considerable expense, we had to buy a twenty-year-old Holden station wagon that was only marginally roadworthy. A critical back window didn't work, the oil gauge was broken, one door would only sometimes open, and our gas mileage was less than half of what we'd been getting in the van. The junkyard loss of the van, and the subsequent substantial loss upon resale of the Holden before leaving, accounted for more than half our total expenses for the entire field season.

Nancy spent scores of nerve-racking hours with the five hundred or so finches that we caught in the course of our stay. She continually reassessed which birds to keep and which to release, trying to select those to take by using several criteria: age, bill color, tail length, feather quality, and (as best she could determine) sex.

I had one brief visit with Heidi at the Conservation Commission office in Darwin shortly after my arrival. She told me that the Gouldian finch was now listed on three endangered species lists. One was the International Council for Bird Preservation in Cambridge, England. Another was the Royal Australian Ornithological Union. The third was a Commonwealth listing used by the federal government in Australia.

Heidi had trapped only once since the beginning of 1992, and then only at one of her two principal sites—in the Yinberrie Hills gold-mining area north of Katherine. She'd caught no Gouldians. Except for a lone student who had collected small bits of data over a span of two weeks, the Gouldian research on Newry Station was finished. Heidi doubted that she'd ever return to fieldwork on the Gouldians.

Heidi looked and sounded exhausted, older than I'd remembered her. My perceptions were probably colored by her enthusiasm for the Gouldians, which had waned. At the same time, she was under obvious pressure from her bosses to produce tangible results for the many years she'd spent working on the Gouldians.

As we neared the number of longtails we were allowed legally to export, I nagged Nancy to confirm, and then reconfirm using different sources, that the quarantine regulations for handling the finches between Darwin and Sydney didn't have hidden clauses. I also pushed her to double-check with the various airlines, so that no one along the baggage route could suddenly decide that we were lacking necessary papers or had improperly boxed the birds, and they couldn't be accepted or flown to their next destination.

A week before we were to leave for Darwin with the finches, Nancy called her Australian export agent. He told her that we could not send the longtails to the United States. A countrywide quarantine on birds had just gone into effect. A virulent avian influenza had been discovered near Geelong, Victoria, and sixteen thousand chickens had to be killed. After giving her the disastrous news, the agent said that he'd be delighted to help. For the three months he claimed the finches would now have to be in quarantine, he'd house them in Sydney for $A3,000 a month.

Nancy made several calls to the U.S. Department of Agriculture and to the Australian embassy, and a solution was found. Geography had worked in our favor. The influenza outbreak had occurred more than 2,500 kilometers to the south. For the purposes of our export, the Northern Territory would be thought of as a foreign country.

In Darwin, prior to Nancy's and Cole's and the birds' flight home (I stayed behind for research unrelated to birds), we were ordered out of a motel because the birds in our possession were considered a health hazard. A maid had found them in a shower when we briefly stepped out for coffee. That hurdle surmounted, one more remained: that of getting them through a quarantine inspection. When finally arranged, it required all of a two-minute look-see by an inspector through a dirty car window, but an hour of paperwork and payment in cash of $A450 for a signature saying the blackhearts were healthy and free of disease.

When Nancy and Cole left Darwin in late August, they were accompanied by exactly one hundred forty chirping finches, huddled in two small rectangular wooden boxes amply supplied with finch seed and lots of water. Despite the long flight to Sydney, a half-day stay in a hotel closet, and another long flight across the Pacific to Los Angeles, every one of the finches survived. Nor did even one die in subsequent months while in federal quarantine in Irvine and in their spanking new outdoor aviaries at the University of California in Irvine.

Once out of quarantine and in their spacious walk-in flyways, with unlimited food and plenty of nest boxes in which to build feather-lined (Sears' pillows!) nests, the longtails began reproducing. And they continued to do so, though not always on a predictable schedule. As always, there are complications in research. An exotic parasitic nematode found its way into the colony, and its presence has been implicated in the deaths of a number of longtails. Many of the longtails have been run through gentle and undemanding experimental protocols in Nancy's lab. Numerous offspring of the offspring will be subjected to a variety of harmless mate-choice experiments to test hypotheses concerning their sexual attractiveness and willingness to breed with birds with unlike bill colors and other phenotypic distinctions. Once again, the National Science Foundation has found Nancy's research worthy of generous funding.

By the fall of 1999, despite the best efforts of bird parasite experts around the country, the longtails continue to be plagued by a pattern of fitful breeding, quite dissimilar to the way the first generation behaved in its first two years in Irvine. Nancy continues to work on the breeding and parasite problem and does experiments with longtails, but she has also started up a series of breeding experiments with zebra finches. While she has no plans to abandon research on longtails, it has seemed too risky to now give them all of her attention.

Since 1991 Nancy and I have corresponded fitfully with only one of the students, Amy. Other than a brief sighting of Jean in the fall of 1991 at a departmental party, and a couple of encounters between Nancy and Nat before the Illinois lab was permanently closed, we have seen none of the students.

Jean married her boyfriend, Alfred, not long after returning to Illinois. They have two children, and Jean is working for the Illinois Department of Natural Resources.

Shortly after returning to Illinois, Tim and Amy went their separate ways. For brief periods, Amy worked for an environmentally friendly clothing store, then as a waitress in Aspen, Colorado. When we last heard from her she was married—not to Tim—and living in St. Louis. She was employed as a nanny. In seeking a letter of recommendation from Nancy (one gladly given), she had come to see this as her calling in life.

In the winter of 1991–92, Tim applied to three of the very best graduate schools in evolutionary biology in the United States, to pursue a Ph.D. Among these schools was Cornell University, his top choice. Because of his field stint in Australia, he felt compelled to ask Nancy for a letter of recommendation. She wrote one that was candid, honest, and fatal. Tim was admitted to none of the programs to which he applied. The following year, forgoing a letter of recommendation from Nancy but getting a superb reference from Lana Otsinger among others, he was admitted into a doctoral program in ecology and evolution at a major Midwestern university. He has now finished his Ph.D. and is looking to pursue a career in academia.

Nat, who had already spent a year in a Master's program in biology at Illinois when she went to Australia, and despite failing grades from Nancy that should have eliminated her from the program, continued working on the degree upon her return in the fall of 1991. By 1999 Nat still had not finished the M.S. degree under the tutelage of her Princeton-educated adviser, currently chair of Nancy's former home department at the University of Illinois.

Cole, now nine and in the fourth grade, is an avid—nay, obsessed—collector of butterflies, beetles, and other insects. His considerable collections, housed in the classy black boxes used by professional taxonomists, are the envy of his many friends and especially his parents. Although he occasionally works in Nancy's lab and records data on finches, he proudly describes himself as an entomologist. About his commitment there can be no doubt.

References

I. Works Cited in Text

Aitchinson, R. 1972. *Thanks to the Yanks: The Americans and Australia.* Melbourne: Sun Books.

Alomes, S. 1988. *A nation at last? The changing character of Australian nationalism, 1880–1988.* North Ryde, New South Wales: Angus & Robertson.

Andrew, M. H., P. N. Gowland, J. A. Holt, J. J. Mott, and G. R. Strickland. 1985. Constraints to agricultural development: Vegetation and fauna. In *Agro-research for the semi-arid tropics: North-West Australia,* ed. Russel C. Muchow, pp. 93–111. St. Lucia: University of Queensland Press.

Basedow, H. 1932. The possibilities of the Northern Territories of Australia, with special reference to development and migration. Address to Empire Parliamentary Association, London, Mar. 3.

Beeton, R. J. S. 1977. The impact and management of birds on the Ord River development in Western Australia. Master's thesis, Department of Zoology, University of New England, Armidale.

Bolger, A. 1988. The effect of public sector activity on Aborigines in the East Kimberley. Part II: Aboriginal communities in the Kimberley. East Kimberley Working Paper No. 23. Canberra: Centre for Resource and Environmental Studies, Australian National University.

Burley, N. 1981. The evolution of sexual indistinguishability. In *Natural selection and social behavior: Recent research and new theory,* ed. R. D. Alexander and D. W. Tinkle, pp. 121–137. New York: Chiron Press.

Commonwealth of Australia. 1933. *Memorandum by the government of the Commonwealth of Australia on the development and settlement of North Australia.* Canberra: Commonwealth Government Printer.

Conigrave, C. P. 1936. *North Australia.* London: Jonathan Cape.

Crowley, F. K. 1960. *Australia's western third: A history of Western Australia from the first settlements to modern times.* London: MacMillan.

Crowley, F. K. 1974. *A new history of Australia.* Melbourne: Heinemann.

Dames and Moore. 1982. *Environmental review and management programme: Argyle diamond project.* Perth: Dames and Moore.

Department of Regional Development and the North West. 1985. *Kimberley pastoral industry inquiry.* Perth: Department of Regional Development and the North West.

Drewe, R. 1981. Great government bungles in Australia. *Bulletin* Nov. 3: 38–39.

Edine-Brown, O. B., and G. Webb. 1988. Wyndham crocodile farm: Submission in support of the crocodile farm license application in the East Kimberley. Perth: Government of Western Australia. Mimeo.

Gibbs, D. M. H. 1984. *Assessment of the economic impact of the Argyle diamond mine on the East Kimberley region.* Perth: Government of Western Australia.

Gowland, P. N. 1981. *Lake Kununurra wetlands.* Kununurra: Department of Agriculture.

Graham-Taylor, S. 1980. A history of the Ord River scheme: A study in incrementalism. Doctoral dissertation, University of Western Australia.

Harris, J. 1990. Kimberley flora and fauna: The slide towards extinction. *The Wilderness Society.* Perth, Western Australia, Nov. 14–18.

Headon, D., ed. 1991. *North of the Ten Commandments: A collection of Northern Territory literature.* Sydney: Hoddern & Stoughton.

MacCreagh, G. 1985. *White waters and black.* Chicago: University of Chicago Press.

MacIntyre, J. N. 1920. *White Australia: The empty north, the reasons and remedy.* Sydney: W. C. Penfold.

Markus, A., and M. Riklefs, eds. 1985. *Surrender Australia?* Boston: Allen and Unwin.

Moffitt, I. 1972. *The U-Jack Society: An experience of being Australian.* Sydney: Ure Smith.

Pratchett, D. 1990. DDT concentrations in cattle grazing irrigated pastures in the Ord River irrigation area in Western Australia. *Australian Veterinary Journal* 67: 420–424.

Reid, A. 1971. *The Gorton experiment.* Sydney: Shakespeare Head.

Shaw, B. 1980. On the historical emergence of race relations in the Eastern Kimberleys: Change? in *Aborigines of the West: Their past, their present,* ed. R. M. Berndt and C. H. Berndt. Perth: University of Western Australia.

Storr, G. M. 1980. Birds of the Kimberley division, Western Australia. Western Australian Museum Special Publication No. 11.

Symanski, R. 1990. *Outback rambling.* Tucson: University of Arizona Press.

Taylor, R., and W. Burrell. 1981. *Shire of Wyndham-East Kimberley, town of Kununurra, Town Planning Scheme No. 4.* West Perth: R. Taylor and W. Burrell Consultants.

Trengrove, A. 1969. *John Greg Gorton: An informal biography.* Melbourne: Cassell.

Tropical Resource Management Party. 1988. Report on crocodile farming in the Kimberley, for the Department of Regional Development and the North West. Darwin, Northern Territory. Mimeo.

Willard, M. 1967. *History of the White Australia policy to 1920,* 2nd ed. Carlton, Victoria: Melbourne University Press.

Younger, R. M. 1970. *Australia and the Australians: A new concise history.* Adelaide: Rigby.

II. Select Chronological Bibliography of Finch Research by Nancy Burley

Burley, N. 1981. Sex-ratio manipulation and selection for attractiveness. *Science* 211: 721–722.

Burley, N. 1982. Reputed band attractiveness and sex manipulation in zebra finches. *Science* 215: 423–424.

Burley, N., G. Krantzberg, and P. Radman. 1982. Influence of colour-banding on the conspecific preferences of zebras finches. *Animal Behaviour* 30: 444–445.

Burley, N. 1983. The meaning of assortative mating. *Enthology and Sociobiology* 4: 191–203.

Burely, N. 1985. Leg band color and mortality patterns in captive breeding populations of zebra finches. *Auk* 102: 647–651.

Burley, N. 1985. The organization of behavior and the evolution of sexually selected traits. In *Avain Monogamy,* P. A. Gowaty and D. W. Mock, eds. *Ornithological Monographs* 37: 22–44.

Burley, N. 1986. Sexual selection for aesthetic traits in species with biparental care. *American Naturalist* 127: 415–445.

Burley, N. 1986. Sex-ratio manipulation in color-banded populations of zebra finches. *Evolution* 40: 1191–1206.

Burley, N. 1986. Comparison of the band-colour preferences of two species of estrildid finches. *Animal Behaviour* 34: 1732–41.

Burley, N., and C. B. Coopersmith. 1987. Bill color preferences of zebra finches. *Ethology* 76: 133–151.

Burley, N. 1988. Wild zebra finches have band-colour preferences. *Animal Behaviour* 36: 1235–37.

Burley, N. 1988. The differential allocation hypothesis: An experimental test. *American Naturalist* 132: 611–628.

Burley, N., R. A. Zann, S. C. Tideman, and E. B. Male. 1989. Sex ratios of zebra finches. *Emu* 89: 83–92.

Burley, N., and P. J. Bartels. 1990. Phenotypic resemblances of sibling zebra finches. *Animal Behaviour* 39: 174–180.

Burley, N., C. Minor, and C. Strachan. 1990. Social preferences of zebra finches for siblings, cousins and non-kin. *Animal Behaviour* 39: 775–784.

Featherston, I., and N. Burley. 1990. Do zebra finches prefer to mate with sibs? *Behavioral Ecology and Sociobiology* 27: 411–414.

Burley, N., and D. K. Price. 1991. Extra-pair copulation and attractiveness in zebra finches. *Acta XX Congressus Internationalis Ornithologici,* pp. 1367–72.

Burley, N., S. C. Tideman, and K. Hallupka. 1991. Bill colour and parasite loads of zebra finches. In *Bird-Parasite Interactions,* J. E. Loye and M. Zuk, eds., pp. 359–376. Oxford: Oxford University Press.

Burley, N. T., D. K. Price, and R. A. Zann. 1992. Bill color, reproduction, and condition effects in wild and domesticated zebra finches. *Auk* 109: 13–23.

Johnson, K., R. Dalton, and N. T. Burley. 1993. Preferences of female American goldfinches (*Carduelis tristis*) for natural and artificial male traits. *Behavioral Ecology* 4: 138–143.

Price, D. K., and N. T. Burley. 1993. Constraints on the evolution of attractive traits: Genetic (co)variation of zebra finch bill color. *Heredity* 71: 405–412.

Burley, N. T., D. A. Enstrom, and L. Chitwood. 1994. Extra-pair relations in zebra finches: Differential male success results from female tactics. *Animal Behaviour* 48: 1031–41.

Price, D. K., and N. T. Burley. 1994. Constraints on the evolution of attractive traits: Selection in male and female zebra finches. *American Naturalist* 144: 908–934.

Zann, R. A., S. R. Morton, K. R. Jones, and N. T. Burley. 1995. The timing and breeding of zebra finches in relation to rainfall in central Australia. *Emu* 95: 208–222.

Burley, N. T., P. G. Parker, and K. Lundy. 1996. Sexual selection and extra-pair fertilization in a socially monogamous passerine, the zebra finch (*Taeniopygia guttata castanotis*). *Behavioral Ecology* 7: 218–226.

Burley, N. T., and R. Symanski. 1998. "A taste for the beautiful": Latent aesthetic mate preferences for white crests in two species of Australian grass finches. *American Naturalist* 152: 792–802.

Burley, N. T., and J. D. Calkins. 1999. Sex ratios and sexual selection in socially monogamous zebra finches. *Behavioral Ecology.* 10:626–635.

Index

Aborigines: in Tennant Creek, 42–44; living conditions, 43–44, 137–139; on stations, 45–46, 124, 137–138, 148; shooting goats, 63; treatment of on cattle stations, 66, 136–138, 148; in Kununurra, 81, 137–140, 158; working with cattle, 118, 137–138; payment for work, 119; sacred sites, 123; numbers in North, 172

Apostlebirds, 131

Asians: in Northern Australia, 171–175

Asymmetry measurements: theory of, 104–105; and mate choice, 105; and zebra finches, 105–106; accuracy of, 106–107; collecting data on in blackhearts, 188

Auvergne Station, 137

Beak color: measurement of, 26; significance of, 86, 131. *See also* Blackheart finches; Gouldian finches

Birds: abundance in Kimberleys, 130, 153–155; treatment of by farmers, 153–155

Blackheart finches: meaning of, 8, 48–49; effect on Gouldian behavior, 49; prior knowledge of, 50; diet, 54–55, 109, 146–147, 149; objectives for, 55–56, 64–65; data collection, 56, 100, 120, 186–189; interaction with Gouldians, 58; nesting behavior, 58, 87, 167, 187–188; beak color cline, 61, 187–189; behavior at mistnets, 70, 101; numbers at waterholes, 70, 146–147, 202; use of artificial water, 74, 202; predators on, 87; mating behavior, 87, 93–94, 167, 187; banding success, 101, 186; losing birds, 103; maturity of, 109; mobility of, 109–110; observing behavior, 120–121; bib size, 167, 187; importing to the United States, 191–192, 202–205; condition and problems with in the United States, 206; experiments on in the laboratory, 206. *See also* Asymmetry measurements; Beak color; Gouldian finches; Population sizes

Blackheart Hotel, 67, 157–158. *See also* Newry Station

Blind: use of, 120–121

Bowerbirds, 114

Breeding season: determining length of, 78. *See also* Blackheart finches

Brolgas, 153

Broome, 189

Burning, 57, 85

Butcherbirds, 169

213

Cattle: sacrifice areas for, 73; treatment of, 118, 133, 166. *See also* Station help
Cattle stations: overgrazing, 73; food for help, 98–99
Climate. *See* Drought
Conservation Commission of Northern Territory: mission of, 61, 117; conflict with gold miners, 116–117; effectiveness, 132. *See also* Gouldian finches
Corellas. *See* Little corellas
Crocodiles, 51–53
Crows. *See* Wildlife
Cumbungi, 129–130
Curtin Springs Station, 157

Development in Northern Australia, 172–176
Diet, of finches. *See* Blackheart finches; Gouldian finches
Dingo Creek: trapping at, 69, 71–72, 113, 115, 122, 159, 183, 202; Gouldians caught, 76–77, 159. *See also* Blackheart finches; Gouldian finches
DNA: and determining paternity, 17
Donkeys: food for crocodiles, 52
Drought: effect on finch behavior, 110; importance of, 110

Ecological data collection, 103, 148–149, 167. *See also* Blackheart finches; Gouldian finches
Export of birds, 155, 191–192, 202–205. *See also* Blackheart finches
Extra-pair fertilizations, 3, 17, 188; varying rates, 17–18; male-biased view, 18. *See also* DNA

Farming in Ord Valley: use of insecticides, 91–92; power of farmers, 130; and treatment of bird populations, 153–155
Fears: of snakes, 90; of sun, 90–91
Fieldnotes. *See* Students

Field site: naming of, 67; preparation of, 67, 68. *See also* Blackheart finches; Dingo Creek; Newry Station
Fieldwork: mistaken assumptions about, 2; studies of, 6
Finches. *See* Blackheart finches; Gouldian finches; Masked finches; Zebra finches
Fires. *See* Burning
Flinders Ranges. *See* Gammon Ranges
Fluctuating asymmetry. *See* Asymmetry measurements

Gammon Ranges: feral goat population, 63
Geese, 154
Goats: destruction by, 63
Gouldian finches: range of, 48; promotion of, 51, 204; air-sac mite problem, 51, 58, 113–114; filming of, 53–54, 177; research team, 53–54; study sites, 54, 60–61; data collection on, 54–55, 58, 68, 160, 169–170; diet, 54–55; nesting behavior, 54, 58, 188–189; fires and behavior, 57–58, 61; interaction with blackhearts, 58, 169–170, 189; endangered status of, 61, 170, 189, 204; water for, 69, 202; population size, 77, 117, 123, 160; band problems, 113; beak color measurements, 131. *See also* Conservation Commission of Northern Territory; Dingo Creek; Population sizes; Sex ratios

Horses: food for crocodiles, 52

Illness: Cole, 88–89, 160–161; insect bites, 89, 121–122; among students, 108, 133–134, 183; rashes, 121–122, 158–59
Indistinguishability: sexual, 167
Invermectin. *See* Gouldian finches, air-sac mite problem

Jackaroos. *See* Station help
Jillaroos. *See* Station help

Kangaroos: shooting of, 37–38, 62; market for, 41
Kununurra. *See* Aborigines

Lake Argyle: siltation and overgrazing, 59
Lake Kununurra, 129–130
Little corellas: feeding behavior, 153; shooting of, 153–155
Long-tailed finches. *See* Blackheart finches

Masked finches, 131, 147
Mistnets. *See* Traps
Munsell color chips. *See* Beak color

Newry Station: description of, 54, 65–66, 119; hotel description, 65–66; cattle on, 67; relations with manager, 145. *See also* Gouldian finches; Station help; Wallabies

Oondooroo Station, 38
Optimal foraging, 146–148
Ord Valley. *See* Farming in Ord Valley
Outstations, 139. *See also* Aborigines

Philandering: reasons for, 16–17. *See also* Extra-pair fertilizations
Population sizes: estimation model, 77, 169–171. *See also* Blackheart finches; Gouldian finches

Racism. *See* Aborigines
Rape: student view of, 92–93
Rats in the Ord, 178
Recaptures: rate of, 108–110, 159, 188; among zebra finches, 109; and trapping intensity, 109; significance of, 109, 188. *See also* Blackheart finches; Gouldian finches; Traps; Zebra finches
Red-tailed black cockatoos, 153
Research: objectives, 2–3, 6–8, 115, 128; preparations, 4–5; accomplishments, 115–116; conflict of interest, 116–117. *See also* Students
Residents and flythroughs, 109–110
Ringers. *See* Aborigines; Station help

Sex ratios: in Gouldian populations, 77. *See also* Gouldian finches
Sheep: killing of, 63
Sickness. *See* Illness
Snakes: at field site, 83, 164–165
Station help: living quarters, 66; compensation, 119; cook complaints, 119, 135–136, 157; diet, 119; lifestyle, 119–120, 135–137, 157; perception of, 156
Station managers: behavior of, 59, 148–149. *See also* Newry Station
Students: how chosen, 4–5, 12–13, 15–16, 18–22; previous experience, 11, 19; commitments to, 20, 22–25; payment to, 23, 195–196; teamwork, 24; taking responsibility, 25–26, 111–113; training of, 26; field living conditions, 66, 140; lost luggage, 68, 81; money concerns, 71, 76, 100, 120, 160; preparation for fieldwork, 72, 88; work hours, 80; interest in research, 80, 146; fears, 91, 150; loneliness, 92, 140, 151, 159; paying for babysitting, 95, 120; assessment of abilities, 112; assessment of research, 115–116; assignments, 116; vacation for, 126–127, 132–133; stealing, 128, 144, 160–161, 198, 201; deceptions by, 140–142, 160, 179–180, 182–183, 185, 190, 198, 201; compared to other fieldworkers, 156; summarizing their behavior, 194–197. *See also* Illness; Rape; Research
Study sites. *See* Field sites

Territoriality: of finches, 109
Timber Creek, 131
Ti-Tree Station, 45–46
Top End: contrasted with central Australia, 110

Tourism, 130
Traps: walk-in, 27; setting up, 69, 74; problems with, 69–70; predator problems; 71, 147; success, 74–75, 147; frequency of trapping, 77–78

Wallabies: numbers at Newry Station, 117; as farm pests, 178
Weeds. *See* Cumbungi
White Australia Policy, 171–175
Wildlife: illegal shooting of, 132, 153–155, 178
Willaroo Station, 131–132

Xenophobia in Northern Australia, 171–176

Zebra finches: band color, 1, 17–18, 168; mating behavior, 1–2, 168; sex ratios, 1, 3; size of populations, 15; census of, 47; in Alice Springs, 47, 77, 147; historical numbers of, 62; retrapping in Alice Springs, 77–78; plumage condition, 78; mobility of, 109–110; search for, 180–181. *See* Asymmetry measurements